DATE DUE

IEE HISTORY OF TECHNOLOGY SERIES 15

Series Editor: Dr B. Bowers

ELECTRICAL TECHNOLOGY IN MINING

the dawn of a new age

Other volumes in this series:

ELECTRICAL TECHNOLOGY IN MINING

the dawn of a new age

A. V. Jones and R. P. Tarkenter

Peter Peregrinus Ltd. in association with The Science Museum, London

Published by: Peter Peregrinus Ltd., London, United Kingdom

British Library Cataloguing in Publication Data

A CIP catalogue record for this book
is available from the British Library

ISBN 0 86341 199 1

Printed in England by Short Run Press Ltd., Exeter

Contents

Acknowledgments

The authors would like to express their sincere thanks to the very many people who helped to make the publication of this book possible.

Two people must be singled out for a particular debt of gratitude: Richard Keen formerly of the Industrial & Maritime Museum, Cardiff, and Haydn Morgan, past-president of the Institution of Mining Electrical and Mining Mechanical Engineers, for without their active support, advice and encouragement, the road would have been infinitely harder.

Thanks are also due to the HM Mines & Quarries Inspectorate, Cardiff as well as the librarians and staff of the Institution of Electrical Engineers; Institution of Mining Electrical and Mining Mechanical Engineers; North of England Institute of Mining and Mechanical Engineers; University College, Cardiff; and University of Newcastle upon Tyne, together with those of the public libraries at Newport, Cardiff and Risca.

Thanks are extended to the Industrial & Maritime Museum, Cardiff, and Powell Duffryn plc for permission to reproduce their photographs.

Preface

Electricity, both as a means of providing motive power and lighting, was first used in mines in the British coalfield during the 1880s and this development can be seen as a logical extension to the drive for increased mechanisation. It was quickly recognised that its advantages in areas such as ease and efficiency of transmission, ratio of power output to machine size, controlability, etc. were counteracted by the dangers encountered if used down mines where gas was present. All the equipment employed such as lamps, cables, motors, switchgear, etc. were, under various circumstances, capable of producing sufficient heat via sparks, arcing or incandescent filaments to ignite gas. Efforts were therefore centred on the design and production of equipment that could be used safely in mines. However, these developments were by and large conducted in an unco-ordinated manner by individual manufacturers or users, with the Government being reluctant to take a leading role. In spite of the efforts of a number of notable pioneers, the early lead established in Britain was not fully exploited.

With regard to gas, there were many serious technical problems to overcome, but in areas where gas was not a hazard, the move to using electrical equipment was also relatively slow. The exception was perhaps in Scotland, where the advantages of electrically powered coal cutters were recognised and increasingly exploited.

Whatever the problems in the various fields of electrical development, lack of information should not have been a hindrance. It is evident that in mining engineering circles, close links were established between the various groups both at home and abroad, and between these groups information on all matters of mining interest was readily disseminated. Along these channels, up to date information on electrical developments was also transmitted, to which was added an increasing amount of internationally published official statistical data.

**To our wives
Shirley and Norma**

Chapter 1

Introduction

1.1 Genesis of a new industry

1.1.1 Exploiting a national asset

The application of electrical technology in mining can be seen in retrospect as part of the developing expansion of mechanisation in the mining industry in the nineteenth century. The importance of coal to the British economy was highlighted, and one of the problems hinted at, in an editorial in *The Engineer* in 1881:

> 'Upon its quantity, quality and position depends the material resources of empires. Great Britain, without the wonderful development of this material which pertains to her, would be comparatively unknown in the world. America looks to her vast store of coal as the germ of her present, and more, her future greatness. But coal is not to be obtained in the large quantities required for modern industry without great labour and frequently great danger. The weight obtainable at a surface outcrop is small, and hence the necessity for deep mining.'[1]

Mechanisation was seen as the means of increasing coal output, both in the deep mines, where the problems of pumping and winding were greater, and in the older shallower mines, where the coal face retreated ever further from the pit shaft. In some mines, mechanical haulage was substituted for horse labour, as explained by *The Engineer* at a pit in south-east Durham:

> 'as the distance of the coal from the shaft becomes greater . . . It is impractical to at all times sink new shafts or alter the course by which coals are brought to the surface . . . If then, the system of haulage by machine can be substituted cheaply and efficiently, there would be an enormous gain to the coalowner.'[2]

Some idea of the advantage of a mechanised haulage system can be seen from the report[3] of an electrically powered system in the USA, installed to replace a team of mules. Previously, the loaded tubs took thirty minutes to be pulled up the incline out of the mine but, with the electrified system, the same weight of coal was removed in six minutes.

The record of this industry with regard to innovative techniques or methods of operation is by no means poor, and its influence on the wider industrial scene should not be minimised. Professor J.R. Harris judges that, in developing the use of coal, Britain had built up a world lead in coal fuel technology and furnace development and,

> 'There was already by the 18th century, a number of important coal-using processes which could not be used with other fuels. Because the methods of one industry after another could be converted to coal, British technology was given a long-term programme or target, technological optimism was generated and, as a spin off, a battery of related techniques was distributed around British industry.'[4]

One reason for the estabishment of this firm base, founded on coal technology, was the 'fact that the mining of coal was long familiar and that continual improvements had become expected.'[5]

The economic historian A.J. Taylor maintains that this tradition of improvement was a fundamental part of mining development, for,

> 'Whereas in manufacturing industry, invention has increased productivity and reduced costs, in coalmining it has been and is, the indispensable condition of the industry's existence. Without the steam engine, first for the purposes of mine drainage and then for the raising of coal from increasing depths, coalmining could hardly have survived the 19th century; without improved systems of ventilation and underground haulage, its development after 1850 would inevitably have been severely limited[6] . . . by the 1880's, in all but the smallest collieries, the steam engine was in use both above and below ground and its benefits were being felt throughout every coalfield.[7]

It could be argued that greater use could have been made of machines, including those electrically powered, from about 1890 onwards as demand for coal increased. Such argument would ignore the practical difficulties facing the mining engineer, the problems of investment and finance[8] and the shortage of engineers with adequate knowledge of electrical and mining techniques.

On finance, A.J. Taylor observes:

> ' . . . in general, the London Stock Exchange interested itself little in British coalmining and, as a result, even as late as 1914, mining investment retained much of its local and even its personal character.'[9]

He maintains that the predominance of small mining units, both in an entrepreneurial and an operative sense, mitigated against technical experiment and innovation. The smaller firm might have to wait many years before its investment was recouped, especially when the colliery was not only small but old.[10] The result of this lack of planning and investment was that mining remained a labour-intensive industry until well into the twentieth century.

1.1.2 Difficulties of working

The conditions under which coal is extracted from the earth are hostile, to say the least. The working area may be some considerable distance from the pit shaft, where all supplies have to be brought from and then the extracted mineral transported back, along roadways that may be steeply inclined. The floor may heave or the roof collapse and walls move. There may be potentially explosive mixtures of gas. The problems of coal dust always have to be considered, both as a general irritant and as a potential explosive hazard. The problem of water is common to most pits and, in some, the pumps must operate continuously to prevent flooding. The water presents a further problem to the electrical engineer, not only through the risk of short circuit but also because of the corrosive effect on cables and equipment. Having overcome these and other difficulties, the mining engineer may well have to contend with a coal seam that is very narrow, may be faulted and probably inclined. Providing ventilation to this working area, via a labyrinth of roadways and corridors, will mean moving vast quantities of air over distances of many miles. These working conditions, many hundreds of feet below ground, are accepted as normal by the miner, but few other industries in the nineteenth century presented such a daunting challenge to the electrical engineer. He was asked to provide equipment that could be transported into the confines of a mine and, once there, operate safely and reliably, be easily maintained and provide a return on capital investment.[11] The nearest comparison in present day engineering development would be in oil and gas exploration in the North Sea, where the engineer is also working on the frontiers of science and technical knowledge in order to provide an essential energy resource. Like his nineteenth-century predecessor, today's engineer relies wherever he can on established and proven principles but, where current technology proves inadequate, new systems have to be devised. The operation of new techniques and equipment in a hazardous environment involves some considerable risk to life and property. Today it is fully accepted that, if these risks are to be minimised, the problem should be investigated as thoroughly as possible, codes of practice introduced for both the design and operation of plant, and operators and associated staff comprehensively trained. There will be a time gap between the initial trial installations and the formulation of such agreed working rules and design guides, during which time information fed back can be used to modify and improve the system.

Parallels to this general procedure can be seen to have operated in the application of electrical engineering to mines but, in the early days at least, most developments were carried out in an unco-ordinated manner either by individual engineers, or by firms who had both to develop the new systems and 'sell' them to the prospective users.

The wider application of electricity in mining was a topic taken up by an American mining engineer, F.A. Pocock, at a meeting of the American Institute of Mining Engineers in Washington in February 1890.[12] He quoted the words of a British engineer, John Fox Tallis, to a meeting of the South Wales Institute of Engineers in 1888, illustrating how well information is disseminated among mining engineers. When asked the reason for the relatively slow introduction of this new energy source in mining, Tallis said that, as a rule, mine managers were 'not experienced mechnical engineers; neither are they all versed in electrical engineering.' It may be, he said, that there are a number of

'able and practical electrical engineers who could do the work; but those electrical engineers are not conversant with the practical

> routine of colliery work and are, therefore, placed at a disadvantage by not knowing the actual requirements of mining engineers, and the many little difficulties to be met and contended with in underground working; and until mining engineers have acquired the necessary knowledge and confidence in electricity, it is only natural that they will continue to follow the beaten path.'

Pocock thought this 'hit the nail on the head' and cited a case where an electric haulage plant had been installed near Scranton, Pennsylvania.The question was

> 'not whether electricity could do the work, but whether any electrical company knew enough about the practical difficulties to attack and overcome them. That point being settled, the remainder was easy.'

Just over twenty-five years later, the application of this new technology was well established. Few pits in Britain did not employ electricity, either for lighting or to power surface or underground machine drives. By this time, many of the basic problems were well understood, comprehensive regulations had been issued by the Home Office and a great deal of experience had been accumulated by engineers, both in the field and in research establishments in this country and abroad.

1.1.3 The Association of Mining Electrical Engineers

A comprehensive approach to the multitude of different problems was assisted by the formation in Britain in 1909 of the Association of Mining Electrical Engineers.

The aims of the Association were set out in the first Presidential address[13] by William Maurice and were, broadly, to establish a class of professional mining electrical engineer, and further to

- Consider means of minimising the risk attending the application of electricity to mining and to promote the adoption of approved methods and devices tending to increase safety.
- Promote the general advancement of electrical science in its application to mining, to facilitate the exchange of information and ideas on this subject among members of the Association, and generally extend the experience and increasing the efficiency of members.

In spite of undoubted progress during the opening years of the twentieth century, there were nonetheless 'many subjects which still require a great deal of consideration', W.C. Mountain observed in his Presidential address in 1913.[14] Emphasising the aims of the Association, he called for more members to share their practical working experiences and bring forward technical papers on such topics as

- The design of colliery power stations and the relative advantages of gas engines, steam turbines, high and low pressure engines.
- The design of switchboards and the simplification of switching operations, together with the general points in connection with switchboard attendance, the best class of instruments to use and experience with automatic apparatus.
- Consideration of the best class of cable to be used in shafts and underground,

with the method of laying same and the action of gases, vapours and the impure class of water frequently met with.

- Papers dealing with the practical working of various classes of haulage machinery and the horse power necessary on various gradients, particularly acceleration horse power and the load factors; also three-throw and centrifugal pumps, together with information as to the relative costs and upkeep of these two classes of pumps.
- There was also room for discussion on the questions of compressed air versus electric coal cutting, of how to fix the in-bye cables and gate-end boxes, of how best to obtain satisfactory earths, and on the use of electricity to operate heading machines, the results of trials of different classes of such machinery, and the operating costs.

This is a formidable list of topics for discussion and might, to the casual observer, give the impression of a plea for help in areas where little was known or understood. This would in fact be far from the truth, for by that time considerable knowledge and experience had been accumulated. The essentially amateur approach of the earlier pioneers — in many respects similar to the gentlemen scientists of an earlier period — had given way to the new breed of professional electrical engineers. It was to these men, as well as the purely practical working engineer, that the Presidential remarks were directed, in order that members broadly concerned with these topics, including cable, machinery and equipment manufacturers, consulting engineers and those in research, as well as those employed within the colliery should be kept abreast of developments and benefit from the experience of others.

At that time, 1913, membership of the Association stood at 1,083 and included a number of figures prominent in mining engineering and other fields, including: H.W. Clothier, of A.C. Reyrolle & Co., Prof. F.H. Hardwick, of Sheffield University, W.C. Mountain, Consulting Engineer, R. Nelson, H.M. Electrical Inspector of Mines, and Prof. W.M. Thornton, of Armstrong College, Newcastle upon Tyne.

It is interesting to compare the number of members of the Association with those of the older and larger Institution of Mining Engineers, which in 1890, immediately after federation, had 1,239, and in 1910, just after the formation of the AMEE, totalled 3,254. A large percentage of engineers would be members of both professional groups but the relatively high number in the Association emphasises the intense interest there was at that time in the application of electrical techniques in mining.

The Association focused attention on the particular problems connected with the development of electrical engineering applications in mining. Previously, this had been covered in a more general way by a number of allied institutions such as the Institution of Mining Engineers, the Institution of Electrical Engineers, and the Institution of Civil Engineers.

The Institution of Mining Engineers was formed in 1889 from a federation of four of the regional mining institutions, with the others following some time later. A notable exception was the South Wales Institute of Engineers, who stayed outside the Federation. The institutions forming the Federation were:

- Chesterfield and Midland Counties Institution of Engineers. Founded 1871, federated 1889.

- Midland Institute of Mining, Civil and Mechanical Engineers. Founded 1857, federated 1889.
- North of England Institute of Mining & Mechanical Engineers. Founded 1852, federated 1889.
- South Staffordshire & East Worcestershire Institute of Mining Engineers. Founded 1867, federated 1889.
- North Staffordshire Institute of Mining & Mechanical Engineers. Founded 1872, federated 1891.
- The Mining Institute of Scotland. Founded 1878, federated 1893.
- The Manchester Geological & Mining Society. Founded 1838, federated 1904.

A feature of these and other related institutions was their system of regular meetings at which papers would be presented and technical topics discussed. They produced journals or proceedings, recording in detail the topics discussed. This information was often relayed to a wider public through the many trade journals that flourished during the latter part of the nineteenth century and the early years of the twentieth century.

Through such channels, news of the many new developments was spread, in Britain and overseas, to the diverse interested parties — pit owners, managers, engineers, manufacturers, etc. The information they needed covered the cost of installation and maintenance, performance standards and hazards, new developments and innovations, across the following range of equipment: electrical detonation of explosives; signalling; haulage; pumping; lighting, both safety lamps and fixed general lighting, above and below ground; coal cutting and drilling; drives to surface machinery, such as screens, washing plant, etc.; winding; and ventilation.

For these systems to operate safely and efficiently, a new infrastructure had to be erected, because with few exceptions, standard industrial equipment and techniques (themselves in a state of continuous early development) could not be applied to underground working. Thus, special plant and equipment had to be developed for switching, distribution, protection, etc., to guard against physical damage and the effects of gas and dust. In addition, a separate specialist set of Home Office Rules was issued in 1905, and revised in 1910. In 1908 the first HM Electrical Inspector of Mines was appointed.

1.2 The developing need for technical innovation

1.2.1 Early developments
Historians differ as to when man first mined coal. The Chinese are said to have used it as early as 1000 BC[15] and archaeological remains in Britain show that coal was used as a fuel by the Romans, most extensively in the northern military garrisons, especially the forts along Hadrian's Wall. At the Corbridge depot its use is assumed to have been associated with metal working.[16] There is sparse evidence of coal being used in this country during the Dark Ages through to the Norman Conquest and no mention is made of its use in the Domesday Book, although there are references to metal mining.

Coal was dug on the south side of the Firth of Forth in the year 1200,[17] whilst about the same time the monks of Holyrood Abbey were granted tithe of the colliery of Carriden near Blackness and the monks of Newbattle Abbey were granted a colliery and a quarry on the shore of the Forth at Preston.[18] Around 1239, King Henry III granted a charter to the Freeman of Newcastle to dig coal in the Castle Field and the Forth. By the beginning of the fifteenth century, mining activity had spread to both sides of the River Tyne and extended some miles up river.

Most coal from the early workings was used locally. Greater exploitation was limited by the primitive transport of the day, with difficulties increasing during the winter months. The main exception was the coal taken from the Tyne Valley, where foreign traders in the Middle Ages would deliver their goods to the people of Northumberland and Durham in exchange for local produce, such as cloth and fish, and loaded 'sea coal' as ballast in the hope of trading it in the ports to which they would take their more valuable primary cargo, both in Britain and on the continent.[19] The trade with London is perhaps the best recorded and by the end of the fourteenth century, was well established. There was still a preference for burning timber, due to the smoke and fumes given off from coal, but by the end of the sixteenth century, increasing shortage of wood forced both the industrial and domestic consumers in the capital to accept coal as the main fuel. During the years 1575–80, the annual consumption of coal in London is estimated as 12,000 tons; by the end of the seventeenth century, this had risen to 455,000 tons[20] and the sea coal trade flourished until the coming of the railways in the mid nineteenth century, when other coal fields in Britain found it easier to compete for this trade.[21] The American historian, J.U. Nef, estimates that by the middle of the sixteenth century the average annual production in Britain was just under 200,000 tons. Coal extraction had become established in what were to become the principal mining districts in the British coalfield, except the Kent district, where mining was not started until the end of the nineteenth century.

By the end of the seventeenth century, the total British output had risen to some 2,850,000 tons, a position her nearest rival did not reach until the second decade of the nineteenth century.[22] A hundred years later, this output had climbed to a little under 10 million tons — approximately one ton per head of population, or more graphically an amount that, if made into a wall three feet wide and nine feet high, would extend for 1,800 miles.

The steep rise in output during the latter half of the eighteenth century through to around 1700, resulted from the rapid development of the British economy. This general economic growth involved the expansion of existing industries, such as metal, salt and glass production and ship building, as well as the development of new areas of trade as diverse as gunpowder production and soap-making. Increasingly, coal replaced timber as the fuel for a wide range of industrial processes.[23]

During the eighteenth and early nineteenth centuries, the general economic development continued, characterised largely by the increased use of iron.

1.2.2 Working

Extraction techniques in the early days of mining in Britain consisted simply of removing coal from surface outcrops. Then men began to follow the seam into the hillside and establish a *drift*, the tunnelling distance being limited mainly by problems of ventilation and water seepage. By the fourteenth century, these problems were often eased by the use of *adits*, allowing deeper tunnelling; where

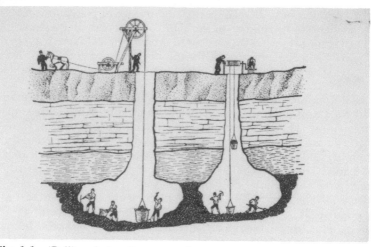

Fig. 1.1   'Bell' coal pits of seventeenth century
(Source: *Coal*, by E.F. Carter, 1963)

geology and terrain allowed, adits were used well into the eighteenth and even the nineteenth century.[25]

More coal could be extracted from the drift by making a series of cuts into the seam at right angles to the main tunnel. In driving these blind tunnels or *stalls*, however, ventilation problems could again arise. These in turn were eased somewhat by linking the stalls with passages running parallel to the main tunnel. This method was later extended, producing a series of interconnected stalls, with coal left between them to support the roof. These remaining sections were termed *pillars,* and the system of working was variously referred to as pillar and stall, board and pillar, post and stall or room and stoop.

The early alternative to drift mining, practised where coal was found a short distance beneath the surface, was to dig vertically down through the overlying strata into the coal to a depth rarely greater than twenty feet. The miner would then open out the base of this shaft by removing coal and creating a cavern, in the shape of a bell (Fig. 1.1.); the size of the *bell pit* would be limited to what was considered safe in that locality. Once this safe limit had been reached, the pit would be abandoned and a further excavation started close by, often resulting in a honeycomb of abandoned diggings.

Gradually, as techniques improved, shafts were sunk to lower depths and by 1700, mines over 100 feet deep were to be found in most districts, whilst on Tyneside, depths of between 300 and 400 feet had been reached. In the last quarter of the eighteenth century shafts on Tyneside were being sunk to below 800 feet and by early in the nineteenth century, mines were being dug to over 1,000 feet. In 1829 the Gosforth Colliery was opened, the lowest level of working being at 1,100 feet.[26]

In sinking these deeper pits, plans had to be made early to provide water drainage, ventilation, the raising and lowering of men and materials, the removal of coal and the method of working the seams: an integrated plan of operation was essential.

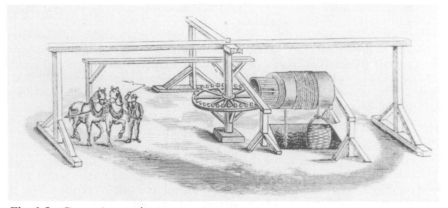

Fig. 1.2 Cog and rung gin
(Source: *Annals of Mining and the Coal Trade*, by R.L. Galloway, 1898)

In Britain, few coal seams run perfectly horizontal. When mining, any water present will follow the inclination of the seam, so the practice developed of drilling the first shaft at the lower end of the mine and, once the pit was working, using this shaft to remove water. Once the working seam was reached, a main road would be driven along the rising seam to some convenient point, or to the outer limit of the proposed workings, where a second shaft would be sunk. In working between these two points, progress would always be 'up the dip', removing coal by the pillar and stall method. The size of the support pillars would depend upon the depth of the seam, the compressive strength of the coal being worked and the bearing strength of the roof and floor. If too much coal was removed, the pillars would either be crushed and the roof come down or, in the case of hard coal, the pillar would be forced into the softer floor and/or roof and again the roof would come down. In either case, the surface damage caused by the land subsidence could be serious and expensive to repair; it could also damage higher seams if these had not already been worked, which in turn could release gas into the current workings.

The amount of coal which should be left behind clearly depended upon a range of factors. In Scotland, in the early part of the nineteenth century, depths of up to 500 feet were worked, but less than two-thirds of the workable coal was removed.

An alternative to the pillar and stall system was *longwall* working, in which a long section of coal face was worked and all the coal removed. As the face moved forwards, the exposed roof was held up by props and the space between the props packed with waste coal and boulders from the workings. The roof would then settle on to this packed area. Roads through this waste (*goaf* or *gob*) were maintained to allow the transport of coal, men and materials.

Other methods of working were, in effect, combinations of these two systems, adopted in different parts of the country depending upon local conditions and, to some extent, established custom and practice.

Longwall working is traditionally said to have originated in Shropshire but was also to be found in various parts of the Midlands in the early eighteenth century. The system slowly spread around the country and about 1760 was introduced to Scotland by miners imported from Shropshire to work pits around Falkirk for the Carron Iron Company.

The longwall system brought with it a number of advantages, in addition to removing all the coal at one time: the production rate of coal was increased by the use of specialist teams of workers, methods of haulage were simplified and smaller seams could be worked. The working face was easier to ventilate too, but this to some extent was counteracted by the difficulties of preventing a build-up of firedamp in the goaf in 'fiery' mines.

1.2.3 Water problems

As deeper mines were sunk in the eighteenth century, water could present serious difficulties. It was not unknown on Tyneside for sinkings to be abandoned due to flooding. The miner would first meet water during the sinking of the pit, as the shaft was dug through water-bearing strata. In large but manageable quantities, the cost of removing this 'make' of water could double or treble the sinking cost; in some cases the shaft would have to be abandoned and another site chosen.

Once the shaft had been sunk, water would present the miner with a two-fold problem: the rate that it had to be removed to prevent flooding and the distance it had to be lifted out of the pit. Various methods were used over the years, from a simple bucket on the end of a rope to an endless chain of buckets, powered by a horse-drawn gin or sometimes by a water wheel.

In Scotland, particularly in Fife, windmills were used to pump water during the middle of the eighteenth century, particularly in the smaller pits where flooding was not too serious a problem and where the sump could collect water on days when there was insufficient wind to power the pumps. There the windmill was more economical than either the horsedrawn gin or the Newcomen engine.

The introduction of the Newcomen engine in the early years of the eighteenth century provided an effective means of removing water in large quantities from deep pits they also allowed deeper mines to be drilled, impossible to drain by earlier pumping methods. Whilst the cost of the Newcomen engine was relatively high (added to which, in the early years, annual royalties had to be paid on the engine), in many cases it was a more economic method of removing water than the horse-driven gin, where to drain large quantities of water, teams of horses might work in shifts round the clock. Horses were expensive to feed and needed trained men to tend to their needs.

By contrast, the steam engine, even with an efficiency of 1% or 2%, was cheap to run, as it could burn the small coal the colliery was unable to sell. In 1713 at the Griff Colliery in Warwickshire, pumping costs of some £900 using horses had been reduced to £150 per year with the employment of a Newcomen engine.[27] The engine installed at the Walker Colliery in Newcastle in 1763 is said to have cost between £4,000 and £5,000 and the total cost of sinking and opening out amounted to some £20,000. Thus, the capital and annual costs of the steam engine, together with any additional cost associated with deeper sinking, had to be balanced against the possible increases in coal production.

By 1775 some 321 Newcomen engines were installed at British pits to pump water and by 1800 up to 1,000 engines of various types, including those by Smeaton and Bolton & Watt, were in operation.[28]

The sinking of new Wearside pits at Monkwearmouth and Murton, illustrates the difficulty that could be encountered in reaching the deeper seams in the first half of the nineteenth century. Work started at the Murton winning in early 1838 of two shafts, each fourteen feet in diameter. A third shaft sixteen feet in diameter was later sunk to a depth of some 440 feet to assist draining feeders. The power used to drain the water from the sinkings came from three pumping engines, and six winding engines converted for pumping. These were powered by steam from thirty-nine boilers and worked twenty-seven columns of pumps. At their peak, the engines were removing 10,000 gallons of water per minute from the shafts. In April 1843, the Hutton seam was found at some 1,490 feet; the sinking had cost between £250,000 and £300,000.[29]

1.2.4 Winding

Increased depth of working also meant greater winding distances and this exposed the limitations of the horse-driven gin. A variety of methods was adopted to increase the rate that coal could be raised, from the use of multi-shafted gins driven by small teams of horses to water wheels. About 1750 an eight-horse gin, with the drum coupled through gearing was in operation at the Walker Colliery in Newcastle. When driven at a rapid trot, they were able to raise six hundredweight of coal 600 feet in two minutes. However, this geared gin was not a success and was removed after a few years.[30]

The water balance became popular in the eighteenth century in the mines of South Wales where adits could be relied upon for drainage. In this system, the power to lift the coal from the pit bottom was provided by the weight of water-filled butts descending under gravity to the foot of the shaft; here the water was allowed to drain away. The empty butts would then return to the surface, again balanced by the empty coal tub returning to the bottom.[31]

If serious restrictions on output were to be avoided, more powerful winding engines were needed and the obvious alternative was to harness steam power. However, rotary motion was required for winding so the power of the early reciprocating beam engine could not be used directly. Steam was used indirectly from about 1760 by taking the water pumped up from the mine by steam to power a water wheel which in turn wound up the coal.

In the 1780s the crank, invented by Pickard, and the sun and planet wheel drive, invented by Watt, provided the impetus to develop the steam engine for winding. In 1784, a Watt engine was used for winding at the Walker Colliery. Although the power of these early engines was limited — rarely exceeding 50 horse power before 1840 — they slowly came into general use. In the small shallow mine the engine was often used for both pumping and winding.

From about 1840, larger high pressure winding engines capable of lifting heavier loads started to be used. At about the same time, wire ropes were introduced into this country from the Continent. These were lighter and stronger than the hemp ropes used previously and better able to raise heavier loads at a faster rate from the new deep mines.[32]

These developments, together with increases in shaft diameters and improvements in shaft guide rails, larger and safer cages, etc., resulted in substantially greater outputs. In the larger pits in the North-East, the winding capacity more than doubled between 1830 and 1850 from around 300 tons per day to some 600 to 800 tons. Further improvements were made in ropes and winding drums and engine output increased. By 1890, a high pressure steam engine installed at Hickleton Main Colliery was able to wind 1,644 feet in 45 seconds and had a maximum performance of 2,763 tons in eleven hours.[33]

In the first decade of the twentieth century, electricity was applied to winding, with a range of drives being developed. As will be shown later, the pioneering work in this area was mainly done by Continental manufacturers.

1.2.5 Haulage

In the early mines, where the drift did not extend too far into the hillside, or where the mine was shallow, the haulage of coal presented few problems. It was either carried in a basket on the back or pulled along the floor in a box or basket.

In the eighteenth century, as pits started to get deeper and economics demanded more extensive working below ground, travel distances increased from the coal face to the shaft and it was not always practical or economic to sink additional shafts following the coal face. On Tyneside, the practice developed of loading coal on to *corves*, pulled on wooden sledges to the foot of the shaft where they were attached to the winding rope and lifted to the surface.

To speed haulage in the extended working, where space and floor permitted, the wheeled corf was introduced. In the early eighteenth century, the flow of coal from working face to shaft was increased still further by the introduction of horses to pull the corves along the main underground roadways. During the latter half of that century, rail tracks were laid along the main roads to the bottom of the shaft and by 1830 this system was well established in Welsh and English collieries.

In Scotland, a similar pattern of development could be seen in the large well-managed pits. The historian B.F. Duckham mentions the installation of wooden plankways from the middle of the eighteenth century in a small number of pits, as well as their use in conjunction with horses on main haulage roads. By the beginning of the nineteenth century, these primitive installations were being replaced by iron rails following the example set in England, principally by John Carr in the Duke of Norfolk's collieries in Sheffield.[34]

In Scotland, as in England, the introduction of new techniques was patchy and the working arrangements of the large go-ahead concerns contrasted sharply with those of small collieries simply catering for the local inhabitants.

In many pits throughout Britain, underground working arrangements were designed around and depended upon child labour. It is therefore not surprising to find widespread evasion of the 1842 Mines Act,[35] which legislated against the employment underground of boys under ten years of age, women and girls. The Act was opposed not only by mine owners but also by many miners and some of the children's parents; economics were the common factor. Evasion of the Act lasted in some pits for many years. Church states that collusion between employers and workers enabled some women to continue working underground for a decade or so, and he estimates that, as late as 1871, some 219 children under the age of nine were still employed underground. This figure, probably an underestimate, compares with about 5,000 children aged between five and ten in 1841.[36]

The loss of cheap labour resulting from the 1842 Act, coupled with the increasing need for greater output, led to the demand for better haulage techniques. However, mechanised face systems were adopted very slowly and muscle power was largely relied upon to transport coal to the main roadway or, in the smaller pit, to the shaft. Boys and small ponies were used in the smaller confines of narrow seams. Galloway records that Shetland ponies were imported to pits in Durham in 1843 for this purpose.[37]

At the South Hetton and Murton Collieries in 1850, there were 214 animals: 96 horses, 37 large ponies and 81 small ponies. By 1871, this had increased to 319: 96 horses, 48 large and 175 small ponies.[38]

This combination of human and horse labour proved flexible and economic, requiring little capital outlay as compared with a mechanical system, although mine owners fully understood that horses would require feeding and looking after, whether they worked or not, and if the horses were to give of their best, they would need to be kept fit and well-fed. This point was brought out by W.B. Brain in a paper read before the South Wales Institute of Engineers in 1882. In his view, significant savings could be made if the capital expended on the purchase of horses were to be spent on electrical machinery, with small unsaleable coal providing steam to drive the generator.[39]

From the 1840s onwards, efforts were made to develop mechanical haulage systems. These extended the principle successfully demonstrated by the steam-driven railway engine, then spreading its network over the countryside, with coal providing both the motive power, and a very useful trading commodity.

By the 1840s steam power had been widely applied to haulage on dip slopes, sometimes complementing the self-acting plane. This system, patented by Michael Menzies in 1750, depended upon the coal seam being inclined at a reasonable gradient from the coal face down to the foot of the shaft. Parallel tracks laid down the incline carried two sets of trucks connected by a haulage rope which ran round a free-running pulley at the top of the incline. The trucks filled at or near the coal face were allowed to run down the incline under gravity, drawing the empty trucks up the parallel track. An obvious drawback was that the main roadway had to be cut wide enough to take the two tracks, but in the right situation, movement of the coal was fast and, once installed, the only cost involved was that of maintaining track and equipment. This system had been adopted slowly where conditions were favourable.

Where the coal face was lower than the shaft bottom, steam power was applied to pull the filled coal trucks up the slope. Empty trucks were allowed to roll back under gravity, pulling out the haulage rope as they went. In many such cases, the steam engine replaced the horse-driven gin.

One of the earliest installations of a steam drive to a level engine plane was at the Peacock mine in Hyde, Cheshire, around 1841; another was at the Haswell Colliery, Co. Durham.

In 1848, what was thought to be the first main and tail rope haulage system was installed in the North of England at Seaton Delaval, Northumberland.[40] This system, the first to be widely applied, used a single track laid along the underground roadway. Two ropes ran from the winding engine: one to the front of the trucks and the other running parallel to the far end of the track, round a pulley and back to the rearmost truck. To wind in loaded trucks, the main rope would be hauled up, while the empty trucks were pulled back by the tail rope. The development of this system coincided with improvements in wire rope manufacture and it was popular in a number of large pits, particularly in the North-East, where it continued to be used after the superior endless rope system had been developed — presumably the advantages of the new system would not compensate for the cost of changeover.

The advantages of the endless rope system were emphasised in the findings of 'The Tail Rope Committee', published in 1868. This committee, set up by the North of England Institute of Mining and Mechanical Engineers, carried out extensive investigations and experiments to determine the relative advantages and costs of the various systems then available. The result was that most haulage systems installed after about 1870 were of the endless rope type, which employed two sets of tracks, along which ran a continuous wire rope. Trucks were attached or detached at any point by quick release clamps. The system could be adapted to the complex workings of the larger pits, where a number of roadways led from the working face to the winding shaft.

These haulage systems were usually powered by steam engines sited either at the surface or at various levels underground. Occasionally, compressed air was employed, the first recorded use being at the Govan Colliery near Glasgow in 1849. The machinery for this installation was supplied by Randolph Elder & Co. The problems associated with these early power drives encouraged some of the first mining electrical engineers to develop electric motors suitable for powering haulage systems.

1.2.6 Gas

Of all the problems confronting the miner, perhaps the most serious was that of gas and this problem tended to increase with deeper workings. Three distinct types of gas present a danger to the miner — blackdamp (chokedamp), afterdamp and firedamp. Each is deadly in its own way and each requires a different set of safety precautions.

In the early days of drift mining, where there was no positive system of ventilation; the main gas problem was *blackdamp*. This gas, mainly nitrogen and carbon dioxide, is formed naturally by the oxidisation of coal and other mineral or vegetable materials. The exhaled breath of the men or animals in the mine also contributes to it. Blackdamp threatens the miner if ventilation is lost for any reason, as at the Hester Pit, New Hartley, Northumberland, where, in 1862, 204 men and boys lost their lives.[41]

This pit had only one shaft, divided down the middle by timber bratticing to form upcast and downcast shafts. The tragedy occurred when the beam of the pumping engine broke and crashed down the shaft, taking much of the bratticing with it. Both shafts were completely blocked. The miners suffocated before rescue teams could clear away the debris. Following this accident, an Act of Parliament was quickly introduced which made it compulsory for all new pits to have two shafts, at least fifteen yards apart. Existing mines were given time to carry out the new sinking.

The presence of blackdamp could normally be detected by the lowering of the flame of the miner's candle or lamp or the miner's shortage of breath. In the early small mines, this warning would generally allow the miner time to move to a safer area or out of the mine before being overcome. In relatively shallow mines where an additional air shaft had been drilled or a later winding shaft had been put down to follow the coal face, a simple form of natural ventilation helped to reduce the problem of blackdamp. This, however, did not always guarantee safe working, as changes in atmospheric pressure or in wind speed and direction could reduce air movement, allowing stagnant pockets to build up, from which escape could prove difficult.

The critical constituent of *afterdamp* is carbon monoxide, a colourless, odourless but highly poisonous gas found in mines after an explosion or fire. Before the recent introduction of the 'self rescuer' gasmask, it was this gas that killed the trapped miners, rather than the explosion or fire itself.

With deeper mines, *firedamp* presented the main gas threat. The miner's candle or oil lamp might ignite small pockets of gas, causing explosion and fire. The earliest reference to this gas comes around the end of the seventeenth and the beginning of the eighteenth century.[42] At this time, the nature of the gas was unknown. The only way the miner could continue to work in the mine with a naked light was to get rid of this gas by ventilation or deliberate explosion. Sinclair described graphically how the 'fireman', covered in wet sackcloth, would enter the area of the mine suspected of containing gas, creeping forward flat to the floor, carrying a lighted candle on the end of a long pole, held high ahead of him (Fig 1.3). Any small pocket of gas would be ignited and, provided the flames ran along the roof of the mine away from him, the resulting explosion would cause no injury to the fireman or damage to the workings.[43]

Fig. 1.3 The Penitent
(Source: *Underground Life or Mines and Miners*, by L. Simonin, 1869)

Gas challenged the miner to provide adequate ventilation and a form of light that would be safe to use in the presence of firedamp. Both problems demanded the attention of many engineers and scientists over many years before acceptable standards were achieved.

1.2.7 Lighting

The first attempt to provide the miner with a 'safe' source of light was apparently invented by Carlisle Spedding between 1730 and 1750. This instrument, the steel mill, was introduced in the Tyneside collieries about 1760. It consisted of a frame

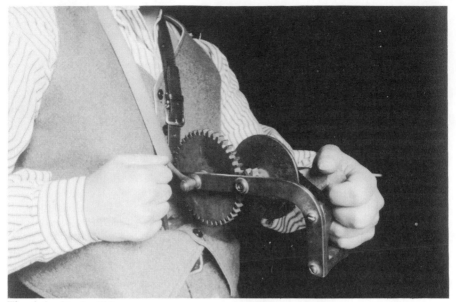

Fig. 1.4   Steel mill in use
(Source: Beamish, North of England Open Air Museum)

housing a toothed wheel about 5 inches in diameter which, when cranked by
hand, engaged and turned a geared shaft (Fig. 1.4). Fixed to this shaft was a thin
steel wheel which, when turned, scraped on a piece of flint held to the wheel by the
miner. A stream of sparks was produced. These sparks gave sufficient light for the
miner to see by, but were meant to have insufficient energy to ignite the gas. The
experienced operator could tell when gas was present by the size and colour of the
sparks: a liquid bluish light showed as the quantity of methane approached the
firing point, blood red indicated chokedamp, or that methane was in excess of the
explosive limit. The limitation of the flint and steel mill was well recognised and it
was also known, on occasions, to have caused explosions.[44] Giving evidence
before the Select Committee of the House of Lords on the State of the Coal Trade
in 1829, John Buddle stated that with gas concentrations of about 1 in 8, the steel
mill could ignite the mixture. He was aware of many such explosions.

Following an explosion at the Felling Colliery, Co. Durham in 1812, when 92
men and boys were killed, a group of local clergy and businessmen, including
John Buddle, set up the 'Society in Sunderland for Preventing Accidents in Coal
Mines'. The main aims were to discover new methods of lighting and ventilating
mines. Buddle stated at an early meeting that nothing further could be done to
prevent explosions by mechanical means and the Society should look to scientific
men for assistance. In 1815 the help of Sir Humphry Davy was enlisted. His first
thoughts were to try to neutralise the gas in some way. He then turned to
producing a light that would be safe in explosive mixtures of firedamp, trying first
of all phosphorous and then a series of sealed lanterns where the air was forced in
or mechanically controlled.

By January 1816 Davy's prototype lantern, shown in Fig. 1.5, had been made

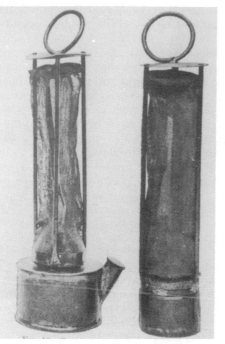

Fig. 1.5 Early forms of miners' safety lamps — the first Davy safety-lamps that were taken underground
(Source: *Transactions IME*, Vol. 51 1915-16)

and successfully tested underground by John Buddle and others. At first it appeared to provide the complete answer to the problem. Davy, however, was at pains to point out that the lamp would not be safe under all conditions found in a working mine. In particular, if the lamp was fitted with only a single wire gauze cover, firedamp could be ignited if the velocity of the gas and air mixture flowing past the lamp was high. In these circumstances, he recommended fitting a tin screen, extended round half to two-thirds of the circumference, and double screens of fine mesh. Sir Humphry Davy and John Buddle carried out tests at the Wallsend Colliery in September 1816 and confirmed these results.

At about the same time, George Stephenson (to become better known as a railway engineer) and Dr William Clanny were also working on safety lamp design and, for some little time after Sir Humphry Davy announced his findings, controversy reigned as to who invented the safety lamp first. None of the three contenders chose to patent their inventions and all had ardent followers and received prizes and commendations for their work. Davy was honoured at home and abroad, and his safety lamp found success on the Continent, being introduced into the Mons region by 1817. By 1825 some 2,000 Davy lamps were in use in the mines of Anzin in northern France.

The safety lamp did not receive the universal acceptance that might have been expected and accidents occurred in spite of its use. John Buddle told the Select Committee on the State of the Coal Trade in 1829, that he thought the introduction of Sir Humphry Davy's safety lamp had reduced the number of accidents from explosions. However, the introduction of the lamp had enabled mines to be worked that were too gassy hitherto and this had increased the danger to a greater degree.

Naked lights were still in use in many British pits in 1835, even where gas could be expected. That year the Select Committee on Accidents in Mines found that the safety lamp gave out too little light and there was difficulty in directing that light on to the working area. That the early lamp was unsatisfactory is illustrated by the large number of lamps invented or modified.[45] Criticism was justified, although it ignored the real progress made in those early days from very primitive beginnings. Some indication of the progress made can be gauged by the comments of J.B. Simpson, in his presidential address to the North of England Institute of Mining and Mechanical Engineers in 1892. He claimed that by 1871 about half the coal mined in Britain could not have been produced without the safety lamp.[46]

1.2.8 Ventilation

Better ventilation was seen as a parallel problem to that of safe lighting. By the beginning of the eighteenth century, pits were starting to be worked with two shafts, to maintain natural ventilation. Increased depth and complexity of underground workings meant greater risks from firedamp and the need for a greater throughput of air. The method chosen at this time at pits with a suitable arrangement of shafts, was to suspend a brazier filled with burning coal down one shaft, referred to as the *upcast shaft*. The resultant updraught would induce a flow of air down the other shaft, referred to as the *downcast shaft*, and thence through the working, carrying any dangerous gases up with the rising warm air and out of the pit. This method of ventilation is claimed to have been used in mines in the Cheadle area of Staffordshire in the 1680s[47] and at the Fatfield Colliery, Co. Durham in 1732.[48] The principle was extended in the latter half of the eighteenth century by the introduction of furnace ventilation. Here, a coal-fired furnace was installed either at the bottom of the upcast shaft, with the shaft acting like a conventional flue, or on the surface with the air from the upcast shaft being directed through the furnace and out through the normal furnace chimney.

In 1760 James Spedding introduced his method of 'air-coursing'. Regulating doors or fixed partitions were installed at key junctions in the elaborate network of passageways, so as to direct the air through all passageways, including the working faces in turn. Manning the doors was a task traditionally carried out by child labour before the 1842 Act and by young boys age ten years or over thereafter. In a large mine the total distance the air had to travel could be in excess of twenty miles. With large travelling distances, the resistance to the air would be high and velocity low, so the air coming from gassy districts could foul sections where gas was not normally present. A later development, called the compound or split system, got round these problems by siting the doors or fixed partitions in such a way that a number of parallel paths were presented to the incoming fresh air. This arrangement enabled larger volumes of air to be introduced (see Table 1.1). The number of parallel circuits used depended on site conditions. Ten to fifteen were not uncommon, with the average length of air courses in the 'great collieries' reduced to less then 3 miles.[49]

The standard of ventilation in the various coalfields between 1800 and 1850

Table 1.1 Ventilation rates at a number of North East collieries *c.* 1850[51]

Colliery	ft^3/min
Hetton	190,000
South Hetton & Murton	132,895
Wallsend	121,360
Haswell	100,917
Tyne Main	94,810
Hebburn	77,005
Monkwearmouth	70,500
Felling	54,000
Walker	44,800
Black Boy	34,000
Coxlodge	20,000
Gosforth	16,000
Mickley	12,856

was very variable. In the early part of the nineteenth century, ventilation rates rarely exceeded 800 ft^3/min. In 1835 the rate at Wallsend Colliery was quoted as being 5,000 ft^3/min. and by 1849, the main air current had increased to 75,000 ft^3/min., travelling at 14 miles/hour, the fastest known at the time.[50]

By 1850 a marked improvement had been made in ventilation rates, at least in some of the North-East pits (Table 1.1). In some parts of the British coalfield, however, even as late as 1849, many pits had no artificial ventilation at all. In south Staffordshire, Worcestershire and Warwickshire, furnace ventilation was scarcely known; here natural ventilation was relied upon, occasionally supplemented by the fire lamp.[52]

Elsewhere, in spite of the inherent risk from explosions, underground fires and damage to the upcast shaft, furnace ventilation was retained throughout the nineteenth century. The underground furnace was particularly popular in the deeper mines of northern England, where some collieries operated with two or more furnaces to common or multiple flues. The figures in Table 1.1 for Hetton Colliery were achieved using three underground furnaces.

Alternative methods of ventilation were tried, including the use of steam jets in the upcast shaft as early as 1811, with more extensive experiments in various parts of the country between 1830 and 1850.[53] From the results available, Galloway concluded that, whilst the steam jet system had its proponents, it did not appear to be as effective or efficient as the furnace.

As early as 1807, Buddle had experimented with a mechanical air pump at Hebburn Colliery. A wooden piston, 5 ft square and having an 8 ft stroke, worked in a wooden boxing at a rate of 20 strokes per minute. The theoretical ventilating rate is quoted at 8,000 ft^3/min (i.e. double acting) but a practical performance of 6,000 ft^3/min. An early fanner machine was installed in Paisley in 1827. This unit, employing large blades, was placed on its side, completely covering the top of the upcast shaft.

From about 1840, the technical press took up the question of mechanical

ventilation of mines but, on the whole, mining engineers in the northern coalfield
at least were sceptical.

A great deal of evidence on the question of ventilation was put before the 1849
Select Committee looking into the best ways of preventing accidents in mines.
Details were given of extensive experiments on furnace ventilation and the use of
steam jets. George Elliot, owner and *viewer* of Monkwearmouth and Usworth
Collieries, Co. Durham, concluded that, whilst there were many situations where
steam jets would be suitable, the system would not be successful to the exclusion
of the furnace. The best application would be a combination of the two.[54]

Some details were also given of recent installations using mechanical
ventilators, including those of Struve and Brunton. In his evidence to the inquiry,
Brunton said that some months previously he had designed and installed a
centrifugal fan at the Gelly Gaer Colliery in Glamorganshire. This, like the unit
installed at Paisley some twenty-two years earlier, was placed over the top of the
upcast shaft. The fan was 22 feet in diameter, with the intake centre section over
the shaft some 8 feet across. Revolving at 95 rpm, it could exhaust air at the rate of
18,000 ft^3/min. The fan, he said, had been installed as an experiment at a small
pit where there was no risk of gas. The owner, satisfied as to its performance,
intended to move it to another mine for permanent installation.[55]

A report on ventilation in mines (recording details of installations in the
North-East, Yorkshire and Derbyshire) was published in 1850. This was
followed by a Select Committee on accidents in mines in 1853 and 1854.

In the 1854 enquiry, Samuel Dobson, the Principal Manager of the Duffryn
Collieries gave an account of the installation of a steam-driven Struve mechanical
ventilator at Middle Duffryn Colliery, following an explosion when 65 miners
were killed. The accident is assumed to have been caused by gas being lighted at
the underground furnace. For safe working, a ventilation rate of some 60,000
cubic feet per minute was needed and it was found that this could not be provided
using the underground furnace (modified to operate with a *dumb drift* or bypass)
assisted by steam jets. Struve's machine, with a reciprocating action, had two
cylinders each 20 feet in diameter. It achieved the necessary ventilation rate
without difficulty.[56]

In their report the committee, without stating a preferred method,
recommended that 'adequate artificial means of ventilation be provided at all
collieries, and that there shall be at all times a sufficient current of air through the
workings to dilute and render harmless all noxious gases.'[57]

Experiments continued in a number of coalfields with both mechanical pumps
and fans but it was not until the 1860s that fans capable of competing with
furnace ventilation were introduced. Centrifugal fans replaced the fanner and
displacement type, with the Waddle and Schiels being introduced in the 1860s,
followed by the more successful Guibal fan.

Good and reliable ventilation is of paramount importance in coal mining.
Mining engineers were anxious to ensure that the equipment they installed would
be acceptable in the broad range of conditions found in British coalfields.

By the middle of the nineteenth century, it was clear that the future growth and
development of the mining industry would depend largely upon the ability of
engineers to provide both safe working conditions underground and the broad
range of mechanical equipment necessary to ensure continued increases in
output. Some form of forum to co-ordinate developments was needed, and in

1852 engineers in the northern coalfield came together in Newcastle to form The North of England Institute of Mining and Mechanical Engineers. One of the first tasks the new institution set itself was to investigate the theory and practice of mine ventilation. The committee appointed made regular reports. A further group was formed to compare the main types of ventilators then in use, looking at twelve different installations — six fans and six displacement systems. Their findings were given in the Institute's *Transactions* Vol.30, 1880–81.

Galloway gives details of a number of installations of the 1850s and 1860s, noting the important introduction in 1864 of the Guibal fan, patented in 1862 by M. Guibal of Mons. Within the next ten to twelve years, a further two hundred or so were installed around the country.[58]

In 1888 a joint committee was set up of engineers from North of England Institute of Mining and Mechanical Engineers, the Midland Institute of Mining, Civil and Mechanical Engineers and the South Wales Institute of Engineers[59]. The committee's investigations were restricted, at first, to those collieries where two fans of different type were ventilating the same shaft so that experiments could be made upon both under precisely the same conditions. They used data available from other experiments at home and abroad, particularly the report of the Committee of the Société de l'Industrie Minerale. Their extensive and detailed report took ten years to prepare.

1.2.9 The challenge of electricity

The period covered by the preparation of the reports on mine ventilation spans nearly fifty years, during which the mining industry in Britain saw a large number of innovative techniques introduced, including the new motive power of electricity. The introduction of new ideas coupled with the complex and interdependent systems within a coal mine constantly challenged the mining engineer and frequently presented a serious threat to life and property. The introduction of electricity underground seemed to some too great a risk to contemplate and to many others something of a mixed blessing. The small band of engineers advocating its use in the early days were to some extent visionaries, aware that dangers existed, but believing these dangers could be minimised and that the eventual benefits would be considerable.

1.3 Notes and references

1 *The Engineer*, 1881, Vol. 52, p.45
2 *ibid.*, 1884, Vol. 58, p. 261
3 *The Electrician*, 1887, Vol. 19, p. 388
4 HARRIS, J.R.: 'Skills in coal & British industry in the eighteenth century, *History*, 1976, Vol. 61, No. 202, p. 168
   Harris asserts that the techniques used in the design and operation of furnaces for both ferrous and non-ferrous metal refining, as well as glass manufacture, were largely craft-based. The skills were in the hand and eye of the workman and this expertise was almost 'unanalysable'. This, he claims, is why the use of coal in these processes was successful in this country but found little success abroad, in spite of numerous attempts at 'intelligence gathering'. J.R. Harris is Professor of economics and social history at Birmingham University.
5 *ibid.*, p. 170
6 TAYLOR, A.J.: 'Labour productivity & technical innovation in the British coal industry 1850-1914', *Economic History Review*, August 1961, p. 51

7 *ibid.*, p .58
8 BUXTON, N.K.: *'The Development of the British Coalfield'*, 1978, p. 115
9 TAYLOR, *op. cit.*, p. 64
10 *ibid.*
11 A graphic account of the problems of applying mechanised drilling techniques in American copper mines is given in LANKTON, L.D.: 'The Machine under the Garden: Rock Drills arrive at the Lake Superior Copper Mines, 1868–1883, *Technology & Culture*, 1983, Vol. 24, No. 1
12 *Transactions American Institute of Mining Engineers*, 1889–90, Vol. 18, p. 412
13 *Proceedings Association of Mining Electrical Engineers*, 1909-10, Vol. 1, p. 15
14 *ibid.*, 1912–13, Vol. 4, p. 4
15 HALL, T.Y.: 'On the progress of coal mining industry in China', *Transaction North of England Institute of Mining and Mechanical Engineers* (NEIMME), 1855-56, Vol. 15, p. 67–74
16 WEBSTER, G.: 'Notes on the use of coal in Roman Britain', *The Antiquarian Journal*, 1955, Vol. 35, p. 200
17 CANTRIL, T.C.: *'Coal Mining'*, 1914, p. 3
18 GALLOWAY, R.L.: *'A History of Coal Mining in Great Britain'*, 1882, p. 5
19 NEF, J.U.: *'The Rise of the British Coal Industry'*, 1932, Vol. 1, p. 10
20 *ibid.*, p. 80
21 London's appetite for coal was such that there was a modest increase in the amount of sea coal delivered following the first rail deliveries in 1845. Prior to the establishment of extensive rail networks,deliveries by sea were the only viable means of supplying coal from the Welsh and Scottish coalfields, as well as from the North-East ports
22 HARRIS, *op. cit.*, p. 170
23 NEF, *op. cit.*, Vol. 1, pp. 190–223
24 MITCHELL, B.R., and DEANE, P: *Abstract of British Historical Statistics*, 1962, p. 131
   POLLARD, S: 'A new estimate of British coal production 1705–1850', *Economic History Review*, 1980, Vol. 33
25 *History of the British Coal Industry*, Vol. 2, 1700–1830
   FLINN, N.W.: *The Industrial Revolution*, 1984, p. 111
26 WELLFORD, R.: *A History of the Parish of Gosforth*, 1879, p. 48
   It is of interest to note that when this pit opened, sufficient interest and enthusiasm was generated, and adequate facilities provided, to be able to entertain some 300 to 400 people to a ball at the pit bottom. The local (Coxlodge) brass band provided the music for dancing and general entertainment. Each guest was invited to go to the coal face and cut a commemorative piece of coal
27 BUXTON: *op. cit.*, p. 23
28 FLINN: *op. cit.*, pp. 120–127. HARRIS: *op. cit.*, p. 170
29 GALLOWAY: *op. cit.*, pp. 208–210
30 GALLOWAY: *op. cit.*, p. 111
   FOSTER BROWN, E.O.: 'History of winding', *Historic Review of Coal Mining*, p. 171
31 FLINN: *op. cit.*, p. 100
   MORRIS, J.H., and WILLIAMS, L.J.: *'The South Wales Coal Industry 1841–1875'*, 1958, p. 70
32 GALLOWAY, R.L.: *Annals of Coal Mining and the Coal Trade* (Second series), 1904, pp. 330/333
   FOSTER-BROWN, E.O.: *op. cit.*, pp. 172–176
   PERCY, C.M.: *'Mechanical Equipment of Collieries'*, 1904, pp. 480–496
33 FOSTER-BROWN, E.O.: *op. cit.*, p. 175
34 DUCKHAM BARON, F.: *'A History of The Scottish Coal Industry'*, Vol. 1, 1700–1815, 1970, p. 103
35 *Select Committee on Accidents in Coal Mines*, 1854, IX, QQ 1232–4
36 *The History of the British Coal Industry*, Vol. 3; CHURCH, R.: *Victorian Pre-eminence 1830–1913*, 1986,, pp. 194/199
37 GALLOWAY, *op. cit.*, 1904, p. 344
38 The Feeding of Colliery Horses, *Transactions North of England Institute of Mining and Mechanical Engineers*, Vol. 32, 1882, pp. 92–93
39 *Transactions South Wales Institute of Engineers*, Vol. 13, 1882/83,, p. 277

40 GALLOWAY, *op. cit.*, 1904, p. 343
41 DUCKHAM, HELEN and BARON: *Great Pit Disasters, Great Britain, 1700 to the Present Day*, 1973, pp. 95–114
42 J.C.: *'The Complete Collier'*, 1708, p. 23
43 SINCLAIR, G.: *'A short history of coal mining'*, 1672, reviewed by Prof. H. Briggs in the *Transactions Institution of Mining Engineers*, 1924–25, Vol. 69, p. 132
44 *'Notes on the History of the Safety Lamp'*, *Trans. IME*, 1915–16, p. 554, Vol. 51, *Report of the Select Committee of the House of Lords on the State of the Coal Trade*, February 1830, p. 33
45 HARDWICK and SHEA, *op. cit.*, pp. 548–702
    Select Committee on Accidents in Coal Mines, 1849, 1853, 1854
    Royal Commission on Accidents in Mines, 1886. The committee looked at around 200 lamps, the majority of which were variants of the Davy or Clanny lamps
46 *Trans. NEIMME*, 1891/92, Vol. 41, p. 171
47 CHESTER, H.A.: *Cheadle, Coal Town*, 1981, p. 22
48 SINCLAIR, J.: *Environmental Conditions in Coal Mines*, 1858, p. 4
49 GALLOWAY, *op. cit.*, 1904, p. 259
50 *ibid.*, p. 257
51 Data taken from GALLOWAY: *op. cit.*, 1904, pp. 256–268, where a more extensive description can be found
52 GALLOWAY: *op. cit.*, 1904, p. 265
53 *ibid.*, pp. 269–286. Details of various steam jet installations can be found in the *Report of the Select Committee on Accidents in Coal Mines* 1849, 1853, 1854
54 *Report of the Select Committee of the House of Lords on the State of the Coal Trade*, 1849, QQ3007 and 3008
55 Further description of progress in the installation of mechanical ventilators in South Wales can be found in MORRIS and WILLIAMS: *op. cit.*, 1958, pp. 64–66 and in JONES, A.V.: *Risca: its Industrial and Social Development*, 1980, pp. 29–30
56 *Report of the Select Committee on Accidents in Mines*, 1854, QQ 2879–2976
57 *ibid.*, p. 8 of the report, recommendation No. 2
58 GALLOWAY, *op. cit.*, 1882, p. 255
59 *Trans. IME*, 1898–99, Vol. 17, pp. 482–576

Chapter 2
Early applications of electric lighting

2.1 Limitations of the traditional safety lamp

As shown earlier, inadequate lighting was one of the many difficulties facing the miner, and conditions underground militated against the provision of good illumination. The presence of gas precluded the use of open flame lamps or candles and enclosing the flame severely reduced the light output. If it fell over, sometimes even if only tilted or knocked, the oil-fuelled safety lamp would go out; the lamp had, therefore, to be placed some distance away from the face being worked, thus reducing the available light still further. If the lamp were to go out, it could not be opened and relit.

Usually the light available at a workplace comes both directly from the light source and from reflections off the ceilings, walls, etc. In the mine more or less uniform black surfaces gave little reflectance and poor modelling. Dust, both in the atmosphere and on the lamp itself, added to the obscurity. The quality of the atmosphere further affected the burning qualities of the lamp fuel. In these adverse conditions, oil safety lamps soon became dirty and at the end of the shift could give as little as one-third of the original output.[1] It is hardly surprising that the eye complaint, nystagmus, was common in safety lamp pits.

The provision of adequate lighting below ground would have been much less of a problem but for the presence of firedamp. As mines got deeper during the nineteenth century, ever more gas was found. The scale of the problem was illustrated by Redmayne:

> 'The volume of firedamp given off in a large and well ventilated
> modern colliery frequently amounts to more than two million cubic
> feet in twenty four hours. Deeper shafts and more extensive workings
> brought about the era of big colliery explosions. Repeated disasters in
> pits led to the formation, at Sunderland in 1813, of a Society for the
> Prevention of Accidents in Coal Mines.'[2]

In Chapter 1 this Society was seen engaging the services of Sir Humphry Davy, leading to the introduction of his safety lamp in 1815. Though this lamp and the many variants produced in the following years were an improvement on earlier lighting, the problem was still far from solved.

The journal *Engineering* in 1869 drew attention to the 'awful loss of life which has been occasioned . . . during the past few years by colliery explosions'.[3] The item went on to refer to the 'very miserable light' given out by the safety lamps then in use, commenting that, in many instances, explosions had been caused by the miner opening his lamp in order to get a little extra illumination.

This item in *Engineering* was designed to bring to the attention of their readers, the contents of a letter recently addressed to *The Times* by Henry Bessemer who was convinced that the thorough lighting and ventilating of a mine was purely a question of cost. Bessemer had put forward an idea for lighting that would allow the mine to be 'lighted like a workshop'. His idea, which he had not patented, was to provide lighting from an ordinary gas burner mounted in an iron box, having one side fitted with a bulls eye or formed from thick plate glass. The gas would come from a gas holder on the surface and the air for combustion would be supplied under pressure, to prevent the entry of firedamp. The heat and burnt gases would be vented through a hole in the top of the box.

Engineering suggested that the idea was 'well worthy of careful consideration' but did not comment on how these lanterns would be initially lighted or maintained, or how an explosion could be avoided in the event of a gas supply pipe being fractured by a roof fall or other accident.

In 1881, *The Engineer*[4] concluded, as had *Engineering* some twelve years earlier, that whilst the lamps of George Stephenson, Dr Clanny, Smith, Upton & Roberts, Martin, Ayre, Whitehead, Fyfe, Elsin, Bosy, Gover and many others were 'more or less perfect', none had been:

> 'invented that was absolutely safe . . . and here, as elsewhere, we so often have to record fatal accidents.'

This leading article went on to describe a novel system of lighting mines, put forward by Messrs. Molcra and Cerbrian in California. This system, the practicability of which they said had never been determined, was to fix a brilliant electric light at the entrance to the pit, and to direct and divide the light through the mine by means of a series of lenses and tubes. There was no doubt, *The Engineer* thought, that this could be done on a small scale but on a large installation would probably be useless.

Both these articles, with their somewhat bizarre proposals, show that the limitations of the traditional miner's safety lamp were well recognised and that, at that time, no one could suggest a satisfactory alternative.

Meanwhile, the problems associated with firedamp were further explored by Professor F.A. Abel, then President of the Institute of Chemistry and Chemist to the War Department. He conducted experiments to determine the explosive power of coal dust. It had been suspected for some time that explosions of coal dust could be triggered off by the ignition of the firedamp. Abel confirmed the earlier findings of Galloway that, not only could coal dust promote and extend explosions in mines, but

> 'the proportion of firedamp required to bring dust in a mine into operation as a rapidly burning or exploding agent . . . is below the smallest amount which can be detected in the air of a mine, even by the most experienced observer, with the means at present in use.'[5]

Safer and better lighting was clearly needed underground and was actively sought, but while the choice of luminants was limited to gas or liquid fuelled flames, little improvement was possible in safety, quantity or quality of light.

2.2 Mains-fed lighting installations

The advent of Swan's carbon filament lamp offered a new approach. Professor Tyndall suggested[6] its possible use as a miner's safety lamp at a soirée of telegraphic engineers following an exhibition of Swan's lamp in October 1880. Two sets of trials were carried out the following year under the auspices of the Royal Commissioners on Accidents in Coal Mines. Professor Abel was one of the Commissioners. The first, in July 1881, was at Pleasley Colliery near Mansfield[7], owned by the Stanton Iron Works Co. This pit was chosen by the Commissioners as a modern pit with longwall working and good ventilation from Waddle fans, and also because it was not a fiery pit, it being normally worked with naked lights.

All the electrical apparatus was provided and fixed by R.E. Crompton & Co., although Col. Crompton was ill and did not witness the tests. Two separate systems were meant to be tested, one using a Gramme machine supplying 24 Swan lamps in single or parallel circuits and the other using a Burgin machine supplying 48 lamps connected in three circuits of 16. The Burgin machine, however, arrived late, so the Commissioners only witnessed one test.

The Gramme machine, sited on the surface, was powered by a 5 hp Marshall engine. Two conductors were taken down the upcast shaft (danger from gas was not expected) to a depth of 580 yards and thence through the workings of the mine, a distance of one-third of a mile. Of the 24 lamps tested, 17 were fixed at the pit bottom, in the main roads or in the gates, and seven were enclosed in hand-held lanterns, specially designed by R.E. Crompton & Co.[8] Each lamp had an output of 15 cp, and two men could work at the face using one lamp and have an abundance of light, except when coal had been newly brought down, then each man needed a lamp for himself. This would mean reducing the output of the lamps at the face to 7½ cp each.

The results of the experiments, reported in *Engineering*[9], were 'regarded as entirely satisfactory by the Commissioners present'; the findings may be summarised as follows:

- The cables can be conveniently arranged not to interfere with the working of the roads.
- The portable lanterns employed sufficiently protect the delicate Swan lamps from rough usage and, with ordinary care, a miner can work without risk to the lamp from anything but heavy roof fall.
- The lamps so protected are probably safe in dangerous atmospheres (this could not be certain until tests had been carried out).
- The increased light will enable the coal to be better hand picked into the tubs so that far less useless material will be sent to the surface. This will probably pay the colliery owner for any increased cost in the provision of lighting.
- The increased light will enable the surveyors to make better roof examinations. In many pits this will prevent many accidents from roof falls.

The second set of experiments was conducted at Earnock Colliery near Glasgow. The installation, described in a number of publications,[10] consisted of a Gramme A type machine installed in a surface workshop, some 250 yards from the pit head, driven by an engine used to power a circular saw. Current was taken from the workshop to the pit head through bare copper conductors ⅜ inch thick, carried on porcelain and vulcanite insulators on telegraph poles. Down the shaft, the two main conductors, each of 19 strands of 22 BWG copper wire, were insulated by a thick coating of gutta-percha, covered with tarred tape and protected by a galvanised iron pipe. The cables leading to the various lamps were Nos. 14, 16, 18 or 22 BWG according to length, the total length of circuit being about two miles. The underground cables were supported on wooden uprights.

In all, 22 lamps were connected to the system, 16 fixed at the pit bottom and in roadways and six portable units used at the faces.

Engineering states that the Earnock Colliery was not a fiery mine but all lamps and switches were arranged as if gas were prevalent.[11] Various forms of switches were experimented with and a mercury tilt switch was finally chosen.

The specific type of lantern used is not clear but, according to *Engineering*, the lamps fixed in the galleries were suspended from the roof and protected by strong glass globes, fitted with silvered copper reflectors. The portable lamps were enclosed in very strong glass lanterns, protected by wire guards.

What appears to be this equipment is described by *The Engineer*[12] in a review of the Paris Electrical Exhibition of 1881. This describes the Swan miner's safety lamp but, more interestingly, gives details of the modifications introduced by Jamieson. A mercury tilt switch is described and a screw contact. The essential difference between the Swan miner's safety lamp and the modified version was that Jamieson introduced a fine wire mesh between the innerside of the strong glass globe and the carbon filament lamp, thus using the principle of the Davy lamp as a safeguard in case of accidental breakage. Swan's safety lamp is shown at Fig. 2.1.

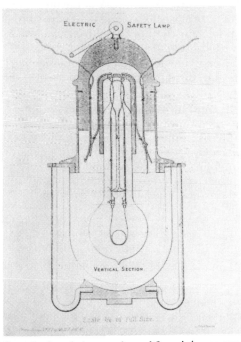

Fig. 2.1 Mr. J.W. Swan's electric lamp adapted for mining purposes
The illustration is taken from the *Transactions of the NEIMME*, 1880–81, Vol 30, plate 38

A brief description is given of a luminaire designed by Edison. In an effort to avoid explosion if the unit was broken, the incandescent carbon filament lamp was protected by a water jacket between the lamp glass and an outer protecting glass globe.

The writer commented that, whilst it was too early to express an opinion on any of these luminaries, their inventions would be of interest to the British people whose wealth so largely depended upon coal and other minerals.

The summary of findings on the tests at Pleasley Colliery, stated that the Swan miner's safety lamp was probably safe in dangerous atmospheres. To check this, a number of lamps were tested to destruction as part of the Earnock trials. In these experiments, Mr. Andrew Jamieson, the Principal of Glasgow College of Science and Art, constructed a box some 15 inches square and 2 feet deep and wired up a Swan carbon filament lamp inside. The box was then filled with coal gas until an explosive point was reached, as tested by a Davy lamp; the Swan lamp was broken and the gas exploded. This clearly established that an electric lamp, if broken in the presence of gas, could be just as dangerous as a naked flame.

The light output from the Swan lamp was found to be considerably higher than that of the Davy lamp. Jamieson[13] found that for a standard 15 cp lamp the average output was 6 cp. By comparison, tests carried out by Dr. Wallace, reported in the *Electrical Review*,[14] found that a common colliery lamp with naked flame had an output of 1·3 cp, whilst that of the Davy lamp was 0·21 cp, the Clanny 0·24 cp and the Teale 0·12 cp. These figures for safety lamps would refer to light output under test conditions as stated previously; conditions in service could reduce the output considerably.

On the question of cost, Jamieson estimated[15] that since the Earnock installation, costs had fallen somewhat and a 50 lamp installation would cost in the region of £350, made up as follows:

Engine fitted into place complete	£100
Dynamo fitted into place complete	£90
One mile of best large leading wire fitted into the mine	£90
50 Swan lamps with protecting lanterns complete	£50
Extras, time, etc.	£15

Jamieson quoted the initial lamp costs as being 12s. 6d., but renewals were only 5s.

The report in the Proceedings of the Mining Institute of Scotland gives the initial cost of the lamps as 25s., with replacements at 5s. each.[16]

A careful check was made of all the lamps installed in the Earnock tests and after some three months it was noted that the minimum lamp life was as low as 10 hours, whilst the maximum was 1,040 hours and still burning. The average life of the lamps that had broken in that period was 280 hours and of those still intact 690 hours.

Luxmore states[17] that the Earnock Colliery was probably the first to use electricity for lighting underground. S.F. Walker, on the other hand, claims[18] that a series of installations for which he was responsible, were the first to employ electric lighting in any colliery successfully. These schemes were at the Nunnery Colliery Company screens outside Sheffield and the Gunner (*sic*) Colliery, Rhondda Fach, followed by one at the Maerdy Pit, also in the Rhondda Fach.

These installations, Walker states, were carried out during late 1879 and 1880. The installation at the Maerdy Pit, at least, involved an element of lighting underground.

Walker's claim implies that the arc lighting installation at the Trafalgar Colliery in the Forest of Dean, which D.G. Tucker[19] says was installed in 1878 to illuminate surface equipment, was not successful. However, from accounts of the electrical pioneering work carried out by the Brain family at the Trafalgar pit, given in Professor Tucker's paper and elsewhere, it seems unlikely that they would fail with an arc lighting system. Arc lighting was relatively well established by then.

The first use of incandescent filament lighting underground appears to be an experimental scheme by J.W. Swan at The Teversall Colliery, Derbyshire, about 1880–81.[20] At these trials, temporary leads were taken down the shaft to test Swan's new lamps at the coal face. For more permanent installations, the Pleasley/Earnock schemes have a strong claim to priority.

The lighting scheme at the Maerdy Pit is worthy of note. Details were given by W. Thomas at a meeting of the South Wales Institute of Engineers in January 1881.[21] In his introduction, Thomas explained that some months earlier, consideration had been given to electric lighting for the surface facilities, pithead and yard, etc., and, after comparing the relative merits of the dynamos and arc lamps then available, a Gramme E type dynamo and Brockie lamp were chosen. The contractor, S.F. Walker from Nottingham, undertook to install the whole electrical apparatus and maintain it for three months at no charge to the colliery company. If, at the end of that period, they were not happy with the installation, he would remove it at his own cost. The colliery had only to provide the motive power for the dynamo.

The rating of this dynamo enabled it to power six lights of the Brockie type chosen, if driven at 1,000 rpm, five at 900 rpm and so on down to one arc light at 400 rpm as required. Selection was provided by a suitable connection to the Waddle ventilating fan which rotated at 40 rpm. A drive belt was fitted to the 16½ foot diameter pulley on the fan drive shaft, then to the dynamo shaft via short counter shaft and a couple of pulleys. Later, a separate small engine was installed to drive the dynamo.

A system of switches and resistances allowed any or all of the lamps to operate at one time and carbons to be changed without risk of electric shock.

As the lighting would not be required on the surface by day, it was decided to install some lighting at the pit bottom. Wires were taken down the pit shaft (there are no details of these cables), which was some 290 yards deep, and connected to two arc lamps 20 yards apart in the roadway that measured 16 feet high by 14 feet wide. Thomas reported that the lamps were operated without covering lanterns and the glare did not inconvenience either the men or the horses.

Arc lamps were chosen for this installation, in preference to incandescent lamps, for their economy. The arc lamps gave some 20,000 candles/hp, whereas incandescent lamps gave only 120 candles/hp.

In 1881, arc lamps were again chosen at Harris's Navigation Collieries, South Wales, for lighting the pit bank and screens. At this colliery, experiments were carried out with both types of electric lamp, the more powerful arc lamp being preferred since, 'On account of the absence of colour in the light, the coal trimmers could pick off 'brass' and 'pyrites' quicker.[22]

The experiments proved conclusively that electric illumination of the surface was desirable at any large mine and that maintenance of the equipment was not a problem, as the ordinary engineering staff could carry out the necessary work after only a week's instruction.

This account also mentioned the introduction of Crompton incandescent lamps down the shaft at the Risca Collieries, South Wales, where they 'gave excellent light and facilitate very greatly the work of the cages and the men in the gate-roads.'[23]

Further evidence that fixed electric lighting was used in gassy mines comes from a description of an installation at the Leycett Colliery of the Madeley Coal & Iron Co., 'admittedly one of the most fiery in the kingdom'.[24] This installation used a number of Edison-Swan 16–20 cp lamps at the pit bottom and to 'within 12 ft of the working face'[25] of a 10 ft thick seam.

A brief outline of two installations is given by S.F. Walker. The first,[26] at Cymmer Colliery, South Wales, had four arc lamps and 40 filament lamps supplied from a Gramme E type machine; the second[27] was at the Eppleton Colliery, County Durham. Walker says that the Eppleton scheme used the then almost universal system, with a compound generator and lamps connected in parallel, so that individual lamps could be turned on or off at will, without altering the speed of the machine or affecting the other lights. The Eppleton Colliery was one of a group of pits in County Durham owned by the Hetton Coal Company, where R.A.S. Redmayne started his career in mining at the age of 18 in 1883. He was to become HM Chief Inspector of Mines (1908–20) and was knighted in 1914.

Another lighting installation in the Durham coal field was successfully carried out by Messrs. Bell Brothers at their Page Bank or South Brancepeth Colliery.[28] They used the system of the Maxim Weston Electric Light Co. There were nine lights at bank, on the pit heap and in the engine houses, and 24 underground; 17 of these were in the Brockwell Pit at the shaft bottom, sidings, stables and engine house, the other seven similarly disposed in the Busty Pit. The dynamo was capable of working 90 lights of 25 cp each. Later the works were extended on two circuits, one operating all night with 37 lamps connected and one all day with 64 lamps in circuit.[29] Maxim incandescent lamps had been used throughout and the current was supplied by two dynamos.

During the next few years, electrical development in mining in Britain was largely confined to lighting installations. These were mainly surface installations but, at a number of pits, were extended below ground by taking cables down the shaft. This was the most common approach to providing a lighting supply underground. Sydney Walker, however, in a letter to T. Connolly about 1886[30] reported that a pit in the Cardiff district had for some three years used an underground dynamo, driven by a Soho engine powered by compressed air,[31] to supply 40 lamps of 20 cp, rated at 48 volts. It appears that there was spare capacity within the surface electric system but, as compressed air was already at the pit bottom, a separate generating plant underground was reckoned more economic than running cables down the shaft. The underground distribution seems to have been of reasonable size for this early period (c. 1883) with the furthest lamp some 500 yards from the dynamo.

An interesting installation at the Cannock Chase Collieries was described by A. Sopwith at a meeting of the British Association in 1886, and variously reported.[32] Sopwith pointed out the convenience and economy of utilising the ventilating fan engine for working the dynamo — it ran at almost constant speed (1–3%). The dynamo was evidently series-wound, for the fan itself acted to some extent as a regulator, checking the dynamo's tendency to run away as lights were switched out. The size of the dynamo was not given, but it took 7–12% of the usual engine output.

Sopwith told his audience that the only original feature of this installation, was the use of old iron and steel pit ropes for main and branch cables. Some four to five miles of cables wore out each year, varying in size from ⅝ to 1½ inches, and the scrap value was only a few pounds per ton. The conductivity of these old cables was about one-seventh that of high conductivity copper cable, but this had not proved a problem. At one of the three installations concerned, current had been conveyed a distance of 13,000 yards (feed and return) and by using a number of cables in parallel, the circuit resistance was only 0.05 ohm, nearly half of which was due to the insertion of a length of ¹%6 high conductivity copper cable. (¹%6 indicated 19 strands of 6 BWG wire).

The routeing and insulation of these old pit cables are equally interesting. Underground, the ropes were simply wrapped with old brattice cloth or tarpaulin, whilst on the surface the ropes were laid in brick channels filled with gas tar and coal dust. Where they were run down the shaft, the cables were encased in wooden boxes, roughly insulated on brackets to avoid the injurious effects of water.

From the brief description given, the scheme appears to have been extensive, taking in the working areas of the mine above ground, which extended over an area of five acres, and included the local church, schools and houses.

From the comments of the author, it appears that the colliery management was not afraid to experiment with various forms of conductors in the name of economy, for he remarked upon the: 'practicability of economical extensions of installations by utilising old materials such as ropes, rails, water and gas mains.'

The system of using old steel or iron wire ropes appears to have been copied widely in the Midlands; by 1891 all pits at Cannock and Rugeley were so fitted.[33]

A further report[34] points out that four dynamos were involved at the Cannock Chase Colliery, three driven by fan engines and one by the saw mill engine. They supplied a total of 333 lamps.

The life of these lamps was prolonged by a factor of two by running them underpowered, thus halving the cost of replacement. The average life of a lamp was over 2,000 hours, and some lasted over 7,000 hours. Brief details were added of a number of 8 cp lamps, fed from accumulators at the coal face.[35]

By 1890, electric lighting was slowly being accepted in the British coalfield. At this time there were no official returns on the number of electrical installations, but in 1889 T.M. Winstanley-Wallis compiled the data in Table 2.1 from information gleaned from District Inspectors. His article, 'Electric Lighting and Transmission of Power in Mining', does not make clear whether these installations were above or below ground.

Contemporary technical literature suggests this list is far from complete, but it gives some idea of the rate that mine owners were adopting electric lighting. At this time, there were some 3,409[37] pits working in Britain; even if the list is out by a factor of 2 or 3, the overall percentage with electric lighting is still rather small.

Table 2.1   List of collieries using electric light[36]

Derbyshire[6]
Babington Company, Cinderhill
Cossall Company, Cossall
Morris and Shaw, Limited, Birch
    Coppice
Sheffield Company, Birley
Stanton Company, Stanton Iron
    Works
Wass and Son, Mill Close Mine

Durham[3]
Hetton Company, Elemore
Elliott and Hunter, Kimblesworth
Earl of Durham, Herrington

Manchester District[3]
Fletcher, Burrows and Co.,
    Atherton
Knowles and Sons, Limited
Exors. of Colonel Hargreaves,
    Burnley

Northumberland[3]
H. Andrews, Esq., Warkworth
Walker Company, Walker, Hilda

North Wales[1]
Foxdale Mine

South Wales[19†]
D. Davis and Sons, 1, 2, 4, 5 Pits,
    Ferndale
Davis and Warner, Bodringallt
Lockett's Merthyr Company,
    Mardy
National Company, National
Watt, Ward and Co., Ynyshyr*
Glamorgan Company, Llwynypia
Albion Company, Albion
Plymouth Works, Abercanaid*
Ocean Company, Park

Nixon's Navigation Company,
    Merthyr Vale
Nixon's Navigation Company,
    Nixon's Navigation
Nixon's Navigation Company,
    Deep Duffryn
Harris' Navigation Co., Harris'
    Navigation Colliery
Glasbrook Bros., Gorseinon*
Thomas, Riches and Co., Clydach*
    Vale
Ebbw Vale Company, Ebbw Vale
Powell Duffryn Company
London and South Wales
    Company, Risca
London and South Wales
    Company, Abercarn

Staffordshire[7]
Cannock Chase Company,
    Cannock Chase
Cannock and Rugeley Company,
    Cannock and Rugeley
Hamstead Company, Hamstead
Madeley Company, Leycett
Hawley and Bridgewood,
    Mossfields
Norton Cannock Company,
    Norton Cannock
Walsall Wood Company, Walsall
    Wood

Wigan, etc.[3]
Nightingale, Potter and Co.,
    Mesne Lee Colliery
Sparrow and Son, Ffrwd Iron
    Works
Wigan Coal and Iron Company

TOTAL   45

* Original printer's spelling errors corrected
† Totals corrected

2.3 Early battery operated lamps

The development of lighting installations during the 1880s was somewhat limited and certainly on nothing like the scale *The Times* would like to have seen — according to a leading article on 27 June 1885. Following a graphic and even lurid description of the recent explosions at Clifton Hall Colliery and the Burnley Pit, it stated that these recent deaths added to the vast total of colliery carnage, calculated at 40,000 lives lost in the previous 30 years, with over one milion men injured. Lack of adequate, safe lighting was pointed to as a major contributory factor in colliery accidents, particularly when men opened their safety lamps to get more light (it was also claimed that men opened their lamps to light their pipes), or if gas came through the passage-ways with the ventilating air at velocities for which the lamp was not suited:

> 'Collieries ought to be lighted in a way to dispense with safety lamps . . .
> 'The feeble glimmer a safety lamp affords, intensifies the manifold hardships of a collier's vocation . . . Electricity is the illuminating medium which can supply the light which the miners want without the flame which endangers them. Difficulties naturally beset its application to a labyrinth of narrow, low, tortuous passages like those of which a colliery is composed. No evidence exists to show that they are insuperable; and no proper zeal has been brought to bear upon the conquest of them.'

Pressing home his attack on the electrical engineers of the day, the writer suggested that:

> '. . . scientific brains [should] exert themselves as keenly for the illumination of a murky, dirty coal pit as for the transformation of a plot of ground in South Kensington into a fairyland.'[38]

In reply to this onslaught, Sir F.A. Abel, in his presidential address to the Society of Arts, suggested that the comments were ill-considered and uninformed: it was one thing to make this type of announcement in 'oracular fashion' but was:

> 'quite another thing to apply electric light with safety, even along main roadways, in mines in which firedamp is prevalent. The writer of those lines would have been less confident in his assertions had he sought sufficient information to teach him that the fracture of a glow lamp or the rupture of a conducting wire in a mine might be as much fraught with danger as the injury of a safety lamp.'[39]

Sir F.A. Abel then claimed that, in spite of all the attendant problems, some progress was being made in the introduction of main road lighting underground. He quoted installations at the Risca Colliery, Monmouthshire, Harris Navigation Colliery, near Pontypridd, and the Earnock Colliery, Hamilton, where results had been satisfactory. If, however,

> 'the miner is to have an electric lamp for lighting up drifts and working places, it must be supplied to him in a self-contained and really portable form, with absolute protection or isolation of the glow

lamp from the surrounding atmosphere and with a store of power sufficient to maintain an efficient light for 10 to 12 hours.'

Joseph Swan applied his invention of the carbon filament lamp to a safety lamp of his own design, which he described to the Literary & Philosophical Society of Newcastle upon Tyne on 14 May 1881[40]. He was about to place this lamp at the disposal of the Accident in Mines Commission to test its practicability for use in coal mines at a pit in Nottinghamshire and at Earnock Colliery near Glasgow. He believed its possible use as a safety lamp would be tested in the laboratory of Professor Abel. Swan envisaged the lamp being supplied from a dynamo via fixed wiring, as tests using batteries had not been successful.

The following year, an improved version of the lamp was described.[41] Both lamp and lantern were smaller and now had a small wooden box of batteries that would keep the lamp alight for over an hour. The light output was between 2 and 3 candles. A box of batteries weighing about 20 lb. should keep the lamp alight for about eight hours. There would still be a risk in fiery mines, as tests in Nottinghamshire proved that breaking the lamp would fire gas.

A much improved version of the lamp was presented to the North of England Institute of Mining and Mechanical Engineers on 12 December 1885[42] and the following year to the British Association at Birmingham. This lamp (patent No. 1999:1886), weighed about 6½ lb. and was claimed to provide an average of one candle for 12 hours or 1¼ candles for nine hours. The power came from a battery, integral with the lamp, consisting of four lead acid secondary cells. When assembled, the battery was leakproof, so the lamp could be used at any angle, enabling the miner to inspect the mine walls and roof better than with a conventional lamp — the 'most important essential duty'[43] Sir Frederick Abel referred to this lamp in his concluding remarks to the Society of Arts. Swan was, he said,

'Universally celebrated for his achievements in the matter of glow lamps [and] was encouraged by the [Royal] Commission in the early days of its existence to give his attention to the adaptation of the glow lamp to the miner's use, and has since patiently pursued the subject.'

He described the latest version of the Swan safety lamp in somewhat guarded tones, implying that, even if considerable progress had been made in the development of the safety lamp, with improvements to the secondary cells and overall safety and performance, there was still a long way to go before all the stringent conditions of the Royal Commission would be satisfied, and the lamp available for everyday use or for mine rescue work. However,

'Even if used only as an auxiliary means of iluminating working places, such lights as those which Swan and others will supply will prove very valuable. A similar lamp of somewhat larger size and comparatively much greater illuminating power may also prove invaluable for exploring purposes, especially after accidents due to outbursts of gas, when the best safety lamp may be of little use, even if they continue to burn.'

In spite of his vehement repudiation of *The Times'* attack, Sir Frederick Abel came to a similar conclusion, on two fundamental points at least:

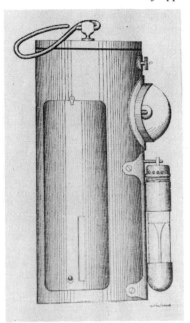

Fig. 2.2   W.J. Swan's improved electric safety lamp
Illustration taken from the *Transactions of the NEIMME*, 1886, Vol. 36, plate 1

- The need for a safe and adequate replacement for the traditional safety lamps.
- Whilst electricity appeared to offer the most appropriate alternative solution, there was at that time no satisfactory electrically-powered safety lamp available for the hard pressed miner.

The leader writer compares the installation at South Kensington with the provision of electric light in fiery mines. Such comparison reveals a total lack of understanding of the physical and technical problems involved. Without this understanding it was unreasonable to expect the dedication or application of the 'scientific brains' to provide answers. The Victorian engineers and scientists were here perhaps victims of their own previous achievements: public confidence in the developing technologies assumed all problems would be quickly solved.

As we have seen, the safety lamp patented by Swan in 1886 was given a fairly wide press.[44] It was a much improved version, incorporating a gas indicator (see Fig. 2.2). This feature was highlighted in a leading article in *The Electrician* in 1887, drawing attention to the need for an acceptable electric safety lamp. The article declared that 'electricity as a means of illuminating mines is steadily making progress'. No doubt, in time, an acceptable lamp would find its way into all mines, but in the meantime:

> 'It rests with electrical engineers themselves and with electrical inventors to press their claims on the attention of the mine owners, and this they should with all the energy at their command.'[45]

To drive the point home, the report of the Chief Inspector of Mines for 1886 was quoted. No fewer than 1,018 lives had been lost out of a total work force of 561,092 and it was fair to assume that 'the greater portion of these casualties was due to either deficiencies of light or the accumulation of explosive gases'.

At this stage the principle of the fixed lighting installation, both above and below ground, was slowly being accepted in British pits, but the lack of a suitable electric safety lamp meant that the miner had still to use the Davy lamp or one of its many derivatives. This continued well into the twentieth century.

For a time it looked as though Swan had made a breakthrough with his safety lamp. A report in *The Electrician* in 1887[46] stated that the Edison & Swan Co. had tried their lamps in the Risca Colliery with success, and Messrs. Watt, Ward & Co. had ordered 2,400 electric safety lamps to supersede all existing lamps in their extensive South Wales Collieries. Only some 600 lamps, however, were actually supplied and put to work, and these were abandoned because in operation the sulphuric acid came into contact with the hardwood casing and generated heat which softened the gutta percha cell walls, releasing more acid and accelerating failure.[47]

A similar fate befell most other lamps introduced during this period. In 1884 S.F. Walker experimented with a lamp using secondary cells but could not resolve the problem of corrosion. He later developed over a period of five or six years, a lamp using zinc/carbon primary cells but this also proved unsuccessful.

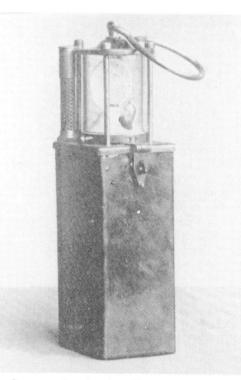

Fig. 2.3   The Sussman electric miners' lamp with gas detector, c. 1900

In A.T. Snell's opinion, the early miner's electric safety lamp was either too heavy and mechanically weak, or too difficult to charge quickly and completely. It was also too costly to maintain.[48]

By 1900 a few electrically powered safety lamps were in general use, these included the Pitkin; Bristol; SCP; and Sussman, which was the most successful.

Sussman, a native of Belgium, first introduced his lamp into Britain in 1893. In 1897, trials were started at the Murton Colliery, Co. Durham, and their success led to some 500 lamps being quickly introduced. In 1899, 1,000 lamps were in use at the pit.[49]

The battery in this lamp was of the Fauré type, consisting of two rectangular ebonite cells, each containing three cast lead grid plates, one positive and two negative. The electrolyte was semi-solid using dilute sulphuric acid. When fully charged, the battery was able to operate for 8 to 10 hours.

Later Sussman lamps were fitted with automatic firedamp detectors. This detector consisted of a perforated metal cylinder, housing a glass tube containing mercury. Finely divided palladium was placed on the outside of the tube and reacted with any firedamp present. The heat of the reaction caused the mercury inside to expand, closing two electrical contacts within the tube and lighting a small red warning lamp. This warning lamp would continue to glow as long as firedamp was present, only going out if the miner moved to a place of safety or the gas was dispersed. This lamp is shown at Fig. 2.3. The gas detector can be seen to the left of the lamp housing.

Fig. 2.4   The CEAG electric miners' safety lamp, *c.* 1912

Initially battery technology limited the development of the electrically powered safety lamp, but this excuse was not valid much beyond 1900. By 1911 there were only 4,298 electric safety lamps in use in Britain; this represented just over a half of one percent of safety lamps in use in that year. This was a marked increase on the 2,055 electric safety lamps of 1910, probably the result of the enquiry into the Whitehaven and Hulton colliery disasters.[50]

The low take-up rate of electrically powered safety lamps was considered serious enough for the Secretary of State for the Home Department to give his backing to a competition for the design that best complied with a set list of requirements. A colliery proprietor offered the prize of £1,000.[51]

Table 2.2   Comparison of different types of safety lamps used in British pits in 1914[52]

(a) **Flame safety lamps**		(b) **Electric safety lamps**	
Lamp	*Number in use*	*Lamp*	*Number in use*
Davy	1,049	Bristol	68
Clanny	193,521	CEAG	44,866
Mueseler	72,044	Float	136
Marsaut	395,139	Gray–Sussmann	8,556
Wolf	17,593	Joel–Fors	491
Hepplewhite Gray	226	Oldham	14,643
	679,572	Thomson–Rothwell	10
		Varta	42
		Wolf	6,895
			75,707

By 1914, the number of electrically powered safety lamps in use in British pits had risen to 75,707 or some 10% of the total (Table 2.2). 59% of the electric lamps were the CEAG type. This lamp, along with the Stach, the Tudor and the Faraday–Hawdon, were demonstrated before a regular meeting of the North of England Institute of Mining and Mechanical Engineers at Newcastle upon Tyne on 14 December 1912[53]. It also showed up well in the competition held in 1911, see p. 182.

The lamp (shown in Fig. 2.4) was of 'extraordinarily strong construction . . . [and could be] thrown onto a stone floor without suffering damage' and had been in practical use with 20,000 miners for a number of years. The outer casing was made from drawn steel plate, heavily tinned and strengthened by pressed corrugations. The battery casing was constructed from strong celluloid, held away from the inside of the steel casing to minimise risk of damage. The top of the battery had:

'a novel acid-proof closing arrangement of the battery cell [which] permits of free egress to the gases without allowing the liquid to escape.'

The top of the lamp was fixed to the lower casing by a bayonet joint and held in place by a strong magnetic lock.

Even with these small battery-operated lamps, there was still a risk of exploding firedamp if the lamp were smashed. To avoid this, the CEAG had the 1.5 cp metal filament lamp housed under a strong glass dome, itself protected by four iron bars. The lamp was held in place and pressed onto electrical contacts under the glass dome by a spiral spring. If the glass dome were broken, the lamp would be automatically disconnected from the circuit. Charging took four to five hours, and this allowed the lamp to burn continuously for twelve to sixteen hours.

A number of advantages were claimed for this lamp, especially:

- A strong steady light that burns with the lamp in any position.
- The user would be less liable to nystagmus or other ophthalmic diseases.
- The lamp had been officially tested and was declared safe against firedamp.

Moreover, 'careful calculations had demonstrated that at collieries with modern CEAG lamp cabins, the cost was no higher than for oil safety lamps.'

In the Newcastle area lamps were often submitted for testing to the mining department at Armstrong College. Prof. H. Louis told members of the Institute that, although the lamp was rated at 1·5 cp, the best he could get from it was 0·986 cp, but that had been maintained continuously throughout the ten hour test.

There was still a considerable following for the traditional safety lamp. In 1913, T.A. Saint thought it worthwhile to carry out an extensive range of tests on thirteen different lamps.[54] In these tests, conducted in the laboratory at the Armstrong College, Newcastle upon Tyne, benzine and various mixtures of mineral and vegetable oil were used as fuel. The best results for light output were obtained with a Wolf lamp fuelled by benzine. With this lamp, an output of 1·0437 cp was obtained at the start of test, falling gradually to 0·974 cp after ten hours. This output compares very favourably with that of the CEAG lamp. The best results obtained for the Marsaut lamp, extensively used in British pits (Table 2.2) varied from 0·7825 cp to 0·66 cp over the test period.

By 1916, the number of electric safety lamps in use in British pits had risen to some 100,000, increasing to 370,000 by 1926.[55] By then the now familiar cap-mounted lamp, with the battery fixed to a waist belt, had been introduced. This was more convenient and provided better light.

2.4 Notes and references

1 *Transactions Institution of Mining Engineers* Vol. 44, p. 274
2 REDMAYNE, Sir R.A.S.: *Men, Mines & Memories*, 1942 p. 159
3 *Engineering*, 1869, Vol. 8, p. 110
4 *The Engineer*, 1881, Vol. 52, p. 45
5 *ibid.*, p. 83
6 *The Telegraphic Journal*, 1881, Vol. 9, p. 379
7 a) *Engineering*, 1881, Vol. 31, p. 621 gives the name of the colliery as being Pleasby
   b) *The Engineer*. 1881, Vol. 51, p. 441
8 *The Engineer*, 1881, Vol. 52, p. 156 says the hand lamps used at the Pleasley tests were Swan lamps modified by R.E. Crompton & Co.
9 *Engineering*, 1881, Vol. 31, p. 622

10 a) *Transactions Mining Institute of Scotland*, 1881/82, Vol. 3, p. 141
    b) *The Electrical Review*, 1882, Vol. 10, p. 52
    c) *Engineering*, 1881, Vol. 32, p. 172
11 *Engineering*, 1881, Vol. 32, p. 196
12 *The Engineer*, 1881, Vol. 52, p. 156
13 *Transactions Mining Institute of Scotland* 1881/2, Vol. 3, p. 145
14 *Electrical Review*, 1882, Vol. 10, p. 52
15 *ibid.*, p. 53
16 *Trans. M.I.S.* 1881/82, Vol. 3, p. 145
17 *Journal Institution of Electrical Engineers* 1979, Vol. 126, p. 869
18 *The Electrical Engineer*, 1896, Vol. 17, p. 21
19 TUCKER, D.G.: 'Early electrical systems in collieries, The Trafalgar Colliery in the
    Forest of Dean and the Brain Family.' *IEE Weekend Paper*, 1975, p. 1
20 *Proceedings Association of Mining Electrical Engineers* 1922/23, Vol. 4, p. 654
21 *Transactions South Wales Institute of Engineers*, 1880–81, Vol. 12, p. 569
22 *The Electrician*, 1883, Vol. 11, p. 274
23 *ibid.*
24 *The Electrician*, 1886/87, Vol. 18, p. 285
25 *ibid.*
26 *Transactions North of England Institute of Mining and Mechanical Engineers*, 1884–85, Vol. 34,
    p. 52
27 *ibid*, p. 54
28 *The Electrician*, 1883, Vol. 10, p. 146, quoting a report in the *Newcastle Chronicle*
29 *ibid.*, p. 242
30 'The progress of electric lighting in mines', *Transactions of Manchester Geological Society*,
    1885/86, Vol. 18, p. 480
31 WALKER, S.F.: *AMEE*, 1912–13, Vol. 4, p. 653. This installation was at the Harris
    Navigation Colliery
32 a) *The Engineer*, 1886, Vol. 62, p. 402
    b) *Engineering*, 1886, Vol. 42, p. 310
33 *Proceedings National Association of Colliery Managers*, 1891, Vol. 3, p. 258
34 *Transactions South Staffordshire & East Worcestershire Institute of Mining Engineers*, 1888,
    Vol. 14, p. 84
35 *ibid.*, p. 90
36 *The Electrical Engineer*, 1889, Vol. 3, p. 510
37 *Reports of Inspectors of Coal Mines*, 1890, HMSO C6346V. Summaries p. 14
38 For more detailed information of this installation see — DUNSHEATH, P.: *A History
    of Electrical Engineering* 1962, p. 148
39 *The Electrician*, 1885, Vol. 16, p. 54
40 *Trans. NEIMME*, 1880/81, Vol. 30, p. 149
41 *ibid.*, 1881/82, Vol. 31, p. 117
42 *ibid.*, 1886, Vol. 36, p. 3
43 PREECE, W.H.: 'Safety lamps in collieries', *The Electrician*, 1888, Vol. 20, p. 379
44 a) *Trans. NEIMME*, 1886, Vol. 36, p. 3
    b) *Report of British Association Meeting*, *The Electrician*, 1886, Vol. 17, p. 359
45 *The Electrician*, 1887, Vol. 19, p. 397
46 *ibid.*, p. 509
47 WALKER, S.F: *Electricity in Mining*, 1907, p. 80. See also WALKER, S.F. 'Electrical
    Miners Safety Lamp' *JIEE* 1900–01, Vol. 30, p. 832
48 JIEE 1900–01, Vol. 30, p. 880
49 *Trans. IME*, 1900/01, Vol. 21, p. 189
50 *Mines Inspectors Reports for 1911*, p. 59
51 *ibid.*, 1910, p. 66
52 Data taken from *Home Office General Report for Mines & Quarries, with Statistics for 1914*, p.
    98
53 *Trans. IME*, 1912–13, Vol. 44, p. 518
54 SAINT, T.A: 'The Lighting efficiency of safety lamps', *Trans. IME*, 1912–13, Vol. 45,
    p. 327
55 *ibid.*, 1944–45, Vol. 104, p. 409

Chapter 3
Early electrical and distribution systems

3.1 Early aspirations

In May 1877 an American mining engineer, N.S. Keith, explained to fellow members of the American Institute of Mining Engineers at Wilkes-Barre, the possible significance of electricity in mining applications.[1] He began by quoting the much publicised statement by Dr. Siemens, that it was perfectly feasible to transmit electrical energy equivalent to 1,000 hp a distance of 30 miles via a copper rod three inches in diameter.

At that time it was not thought possible to run generators in parallel, so, if 1,000 hp was to be transmitted, it would have to be from a single generator. Keith explained to his audience how such a machine might be constructed, with details of the cable to be used and the single 1,000 hp drive motor. He estimated the cost of the three main components and, assuming that the mass of earth could be used as the current return (provided it was of negligible resistance), claimed that at least 50% of the energy expended on the dynamo would be available as mechanical power at the motor.

With the technology then available (the first really practical dynamo had been on the market for barely seven years), there was no hope of such a scheme getting beyond theoretical debate, but it demonstrates that at an early stage the mining engineer, in America at least, was alive to the possible advantages of power distribution by electricity.

Five years later, in 1882, Keith's enthusiasm for electricity had still not diminished. In another paper to the Institution, he drew particular attention to the new applications of electricity:

> '... which can be used by the miner and metallurgist if he has at his main engine or water-wheel a suitable generator of electricity. He may be sinking a shaft at the top of a mountain where it would be impractical to place or operate a steam engine.'[2]

Keith described the possible diverse uses of electricity in and about mines, including lighting, signalling and powering the winding gear, work pumps or rock drills. All of these could be operated by the simple 'turning of a switch'. To back these claims up, he was now able to cite an actual installation — that of the Compagnie de la Ferronière in the Loire Valley, France, where two Gramme machines were being used in a generator and motor mode to raise loads up an incline. This important early installation will be discussed in more depth, together with some of its problems later in the chapter.

Meanwhile, in Great Britain, Sydney Ferris Walker, son of a Devon naval family, was uttering the same sentiments that Keith had done years earlier about the use of electricity in mines. Walker believed that the distribution of power by electricity was:

> ' . . . destined to play an important part in the working of great industries, more particularly of mines.'

He looked forward:

> ' . . . with confidence to the day when the whole of the motive power at a colliery, as well as all its lights and signals, will be furnished by one, or a group of large dynamo-electric machines; . . . When it is remembered that there is no limit in size to electro-motors, it will be seen how great will be the advantages of its universal adoption.'[3]

To illustrate the last claim he said that electro-motors can be made small enough to fit into a greatcoat pocket and large enough to drive a winding engine.

Incredible though such remarks must have seemed to his audience, Walker was in a far better position than most to predict developments. He had been sent by Sir William Thomson (later Lord Kelvin) to inspect Gramme's works in France and was later appointed engineer to the Gramme Magneto Electro Engineering Company. Walker was responsible in 1879–80 for engineering some of the earliest lighting installations in British collieries — Cymmer and Maerdy (both in South Wales) and the Nunnery Colliery, Sheffield.[4]

Although Faraday had demonstrated the principles of electro-magnetic induction as early as 1831, the first practical electro-magnetic machines did not

Fig. 3.1   Cymmer Colliery, Glamorgan, South Wales
Location of one of the first colliery lighting installations, 1879–80

appear until about 1870. Many technical problems contributed to this delay; problems which occupied some of the best minds of the time.

The Belgian, Zenobe Théophile Gramme, eventually produced the first practical machine, described by Professor Tyndall as being of 'exceeding beauty and exceeding power'.[5] One of the reasons for its success, apart from its robustness, was the use of the ring armature, which enabled the first truly continuous direct current to be produced. The armature was wound in such a manner that one or more coils was always generating current as they passed over the space between the poles. Walker, in 1882, despite his connections with Gramme, referred to this generator as the 'Paccinotti machine',[6] a direct reference to the fact that the ring armature was invented by Dr. Antonio Paccinotti of Florence in 1860. To leave no doubt as to whom the machine 'belonged', he described the commutator as 'the Paccinotti commercially developed'.

Whoever devised it, the 'Gramme' was here to stay and was instrumental in the success of arc lighting at the Gard du Nord (1875), the Grands Magasins du Louvre (1877), both in Paris, and the Gaiety Theatre in London (1878). These schemes did much to stimulate the public's interest.[7] Although the German company of Siemens & Halske were producing dynamos for use in telegraphy and firing in mines, Gramme's success prompted the German firm to design and produce powerful and efficient generators for both power and lighting installations.[8] Table 3.1 illustrates the differences which MacKechnie Jarvis, writing in 1958, highlighted between dynamos produced by these two manufacturers in 1873.[9]

Table 3.1   Comparison of Gramme and Siemens dynamos (1873)

	Type	Speed (rpm)	Approx. output (kW)	Weight (cwt)	Cost (£)
Gramme	dc	420	3·2	25	320
Siemens	dc	480	5·5	11	265

Although Siemens was well aware in 1872 that a dynamo was capable of running either as a motor or a generator, several years were to elapse before they could exploit this.[10] The first application was in a mining environment, when Krug von Nidda, Head of the Prussian State Mines, expressed an interest in using electrically operated drills in mines since his compressed air drills were inefficient (von Nidda estimated that only 25% of the power was usable). Possibly because of this and Siemens own mining interests (they purchased a copper mine in the Caucasus in 1863), Siemens embarked upon an extensive programme between 1877 and 1882 exploring the use of electricity in mining. The first fruits of their endeavours were seen in 1879 when they patented a solenoid-operated rock drill. Within two years they had produced a rotary drill. Also in 1879, at the Berlin Trade Exhibition, they unveiled their first electric locomotive designed for use in a mine tunnel, and three years later had installed a complete railway system in a Saxony coal mine.

It was not until 1882 that electricity was used in a power application in a British colliery, and then the machine was of Siemens manufacture.

3.2 The La Perronière system

In his 1882 paper, Keith described the La Perronière system. It had a Gramme machine as the generator, supplying power to a second machine, used as the motor, some 1,200 metres away. The second machine was sited at the head of a 100 metre long incline, having a 1 : 25 gradient. The load raised was 800 kg, in an ascent time of 90 seconds.

An Abstract in the *Proceedings of the Institution of Civil Engineers* and a report in the *Electrical Review* gave a much fuller account of the scheme.[11] These referred to duplicate sets of similar four-pole, series-connected Gramme machines; two as dynamos and two as motors, which were connected by four conductors, forming two independent circuits. Their drives were connected through an arrangement of rollers and pulleys, so that the load could be taken by either or both systems, although one was generally adequate. Each of the cables connecting the dynamos to the motors consisted of 16 strands of pure copper wire, each 1·1 mm in diameter. The cables were insulated with first a covering of paraffined cotton cloth, then two coverings of cloth treated with Chatterton's compound,[12] and finally india-rubber and tar. Although satisfactory under dry conditions, premature cable failure was experienced in damp environments. In an effort to overcome the problem of moisture penetration, lead was applied to sections susceptible to damp, but this led to even more spectacular failures, with lead being splattered against the walls of the pit. The cable manufacturers eventually came up with a formidable combination of gutta-percha and various impregnated cloths to improve insulation under such conditions. The users, not convinced that this was sufficient, further coated the cables with a paste made from Norwegian tar, resin and suet!

The final cost of insulating the lengths of cable susceptible to damp was estimated to be 3·0 francs per metre, compared with 1·25 francs for the original cable.[13]

The result of tests carried out on the system by hauling up the incline first one, then two, three and finally four tubs returned overall efficiencies ranging between 12·2% and 26·1%. These figures, the *Electrical Review* emphasied, were very low but represented the overall system efficiency, including losses in the steam engine, intermediate transmission, friction on the wheels of the tubs etc.[14] Discounting steam engine and transmission losses, the efficiency was estimated to be 37%. The report concludes:

> 'It would be still higher and would amount to about 50% if the work affected were measured, not by the weight of the coal raised but directly on the axle of the motor.'

Clearly, the *Electrical Review* thought it important that there should be no misunderstanding on this point. A figure for true electrical efficiency should be
■ quoted.

The conclusions arrived at by M.M. Charousset and Bague, the French engineers who carried out the tests, as quoted in the *Electrical Review* were:

> 'From their observations that electricity employed to transmit force to a distance will replace with advantage, as regards the results

obtained, the cost of installation and especially of maintenance, by compressed air and mechanical traction, especially in the following cases:

(i)   When the mine is not too much impregnated with fire-damp.
(ii) When the distance between motor and receiver is great.
(iii) When the galleries destined to receive the medium of transmission cables, pipes or chains, are winding, when only one series of galleries and false shafts meeting at right angles are at our disposal for transmission.'

The experiments made at a distance of three kilometres showed that the results varied little with distance. This was also confirmed by similar experiments at another French mine, Blanzy.[15] Here the distance between the dynamo and motor was some 634 metres, 604 metres of which was covered by cable comprising nine strands of 1·1 mm diameter copper wire. The insulation was cloth, with india-rubber and lead for the outer sheaths. The remaining 30 metres of cable was of 5 mm diameter brass wire. The result showed that, with varying loads, efficiency was low when the load was small but increased as the system reached its design value. Thus, at a minimum load of 0·73 hp the efficiency was 17·25%, but at 2 hp the efficiency rose to 51%. Similar tests carried out with the full and short lengths of conductor, i.e. 634 metres and 30 metres, respectively, and had 'no sensible influence upon the efficiency'.

The result of these two sets of tests in France was very important. They demonstrated that electrical energy could be transmitted at relatively high efficiencies over moderate distances, and that, with certain reservations, the use of electricity in mining was feasible both from a technical and commercial standpoint.

In the same year at the Munich Electrical Exhibition a demonstration by Marcel Deprez, a professor at the Paris Conservatoire des Arts et Métiers, and Oskar von Miller, an engineer in the Bavarian Civil Service, added further weight to such claims. They succeeded in transmitting the power from a 1·5 kW steam-driven dynamo at a coal mine at Miesbach along 35 miles of telegraph wire to drive a centrifugal pump which supplied a waterfall at the Exhibition.[16]

The wide reporting of such results on both sides of the Atlantic gave mine owners the opportunity to consider electricity as an alternative form of motive power to extract coal more efficiently.

3.3 Trafalgar Colliery installation

Whilst these developments were taking place on the Continent, similar experiments were being conducted by W.B. Brain at the Trafalgar Colliery in the Forest of Dean. Encouraging results led him to install in December 1882 what was later claimed to be the first application of electricity as a motive power in a mine.[17]

In a paper to the South Wales Institute of Engineers, Brain indicated that his interest in the possible use of electricity in mining went back some years. He told his audience that Professor Silvanus P. Thompson thought that an efficiency of 80% was possible, and Professors Perry and Ayrton claimed to have achieved efficiencies of 60% with comparatively small motors. With larger dynamos and motors the results should be even more favourable. The consensus of opinion was that a maximum of 50% of the work expended in driving the generator could be regained from the driving motor. Therefore, he thought that:

> '. . . the possibility of regaining a percentage power as high as this minimum figure at any point in the mine, regardless of the distance from the surface is of itself sufficient to be worth the attention of those not only engaged in coal and iron but in metalliferous mining also.'[18]

This new motive force would further commend itself where engine coal was of comparatively little value if the same amount of capital was spent on machinery instead of on horses and the small coal was substituted for hay or corn.

Credit must be given to Brain for wanting to carry out his own tests on the efficiency and application of electricity, instead of relying upon press reports of the French installations, although he would have taken note of them. In 1875 he purchased one of the first Siemens dynamo generators in Britain and shortly after obtained the first imported Schuckert dynamo. He then began an involved series of experiments to determine the highest amount of power that an electric motor of a given size and weight could produce. Two machines were specially constructed in the form considered best for mining purposes, combining what were considered the best principles. Additionally, he obtained six Gramme machines, one large E, one B, two A and two M sizes.

In order to perform detailed and meaningful experiments, a test rig was constructed, consisting of a vertical 12 hp engine connected to the boilers of the winding engine and to the necessary pulleys and shafting. These were so arranged as to enable any of the machines to be driven at various speeds, or two or three together. Colliery tubs, filled with iron castings to simulate loads of coal, were pulled up a specially constructed 1 : 6 incline on ordinary colliery rails. The rate of winding, over a fixed distance, was accurately timed. Repeated trials over several weeks encouraged Brain to apply the motor to a pumping arrangement and conduct further tests. The first series involved the surface use of centrifugal pumps and low heads, for which he considered the high speed motor especially suited. He then repeated the tests, with the pumps constructed for high lifts. It was only after these tests had been satisfactorily completed that Brain considered a practical application — that of pumping water from his mine.

The shaft at the Trafalgar was some 200 yards deep and the portion of the mine to be drained was a hold in a rocky seam of coal 500 yards from and 30 yards lower than the shaft bottom, midway between two inclined planes. Although the quantity of water collecting there was not large, it created a problem by running into the rest of the workings 1,000 yards beyond and 200 feet lower. This water-filled hole was difficult to reach to install any other form of power drive. Brain had earlier tried a Warner horse pump but this was unsuccessful because of the difficulty in keeping the water level under control. Air pumps were also considered, but rejected because the cost of the compressor alone exceeded that of a complete electrical installation. The scheme, as finally installed in December 1882, was carried out by the Pyramid Electric Company, under the supervision of A. le Neve Foster.[19] It consisted of a double-acting pump powered by a motor capable of delivering some 4 hp. Problems were initially experienced with the mechanical joints connecting the discharge pipe to the motor which kept breaking down and scattering water about. To protect the pump from the ingress of water, it was moved some distance away, but this incurred a loss of efficiency as a belt drive had to be introduced. Two Gramme type 'A' machines were installed at the surface to provide the motive power, either one of which was capable of carrying the load. An Ayrton and Perry voltmeter and an ammeter were located near the engineman, who also controlled the pump.

1,000 yards of cabling was used in the installation, consisting of 19 strands, each 0·072 inches in diameter. Although the cable was insulated with india-rubber and sheathed with lead pipe, further protection was deemed necessary in the form of a 3 in × 4 in well-creosoted wooden box, secured to the side of the shaft and the roof of the road into the pit.[20] The return conductor appears to have been an ordinary iron rope buried in the ground.[21] Lighting, from the same supply, was provided by Edison lamps at the pump and in the engine room.

Brain's design and installation was clearly superior to that of the earlier French systems, largely because of his professional approach. He insisted upon detailed and protracted testing and so thoroughly understood the system's capabilities before installing it. Regrettably, not everyone adopted this attitude.

The success of the installation encouraged Brain to speculate on possible future uses of electricity at the colliery. He hoped that he would soon be able to describe to fellow members the successful application of electric power to coal getting, rock boring, etc. He also anticipated that in future he would use cables sheathed with galvanised iron wire or, for larger sizes, copper wire. The outer casing would then provide a path for the return current. Looking even further ahead, he expected to describe a 12·5 hp pumping arrangement, together with a similar arrangement for winding. During the subsequent discussion, Brain predicted that the time was not far distant when electricity would be used as a source of motive power 'with perfect immunity from any danger from sparks', and, that when transmitting large quantities of power, with efficiencies of 70%.

Brain's predictions, in part at least, were realised in 1886, when his system was extended by the addition of a 22 kW pump driven by a 320 V dc electric motor. This motor, like the generator, was supplied by Ellwell–Parker of Wolverhampton.[22] The system, which coped with the main flow of water in the deep workings, used 2,000 yards of 19/16 copper wire. The return conductor, however, was a four inch diameter rope, not sheathed as suggested earlier. Initial problems were experienced with the cable insulation enclosed in wooden pipes in the shaft, but a change to lead-covered cable solved these.

Still comparing relative performances, Brain estimated that the useful work done on the water was 35% of the initial steam power. Electrical losses were put at 42%, of which, excluding the generator, the loss of 16% in the cables was the greatest single factor. The steam engine, a second-hand marine type, was in poor condition, and Brain attributed to this losses of 22%. Even so, he regarded electricity as the most convenient means of pumping, as well as being cheaper and quicker to install.

3.4 Subsequent developments

Although papers like Brain's and those from the Continent, along with authoritative contributions to learned societies by prominent, practising electrical engineers, such as Albion T. Snell, Sydney Ferris Walker, W.C. Mountain and D. Selby Bigge, created considerable interest, the anticipated growth in electric power for mines was slow to materialise. There were a number of reasons for this,[23] including the restrictive clauses in the 1882 Electric Lighting Act (see chapter 8); but technology was not sufficiently advanced to ensure safe working underground, especially in fiery mines, and colliery-owners were reluctant to invest in what was, essentially, prototype equipment when the country was in the grip of an economic recession. The figures produced in 1889 (see Table 2.1) showing 3 or 4 power installations in mines and only 45 with lighting,[24] reveal the general attitude.

One of these power installations was at St. John's Colliery, Normanton, where the compressed air installation for pumping was replaced by electrical plant in 1887. This was significant as electrification had previously only replaced horse labour. Recognising this, engineers from many parts of the country came to inspect the installation,[25] which was the largest to be installed in Britain, as *The Electrical Engineer* was quick to inform its readers:

> 'We have several times referred to the use of electricity in mines for pumping purposes . . . but all previous work has been insignificant beside the pumping plant which Messrs. Immisch & Co. have now laid down at Messrs. Locke & Co., St. John's Colliery, Normanton.'[26]

The initial installation at St. John's, started in mid 1887, provided electric power to a small pump delivering 39 gallons of water per minute through a 530 foot head.[27] The results encouraged the mine management to order a larger installation, capable of delivering 120 gallons per minute through a 900 ft vertical head. Albion Snell, Chief Engineer of Messrs. Immisch & Co., undertook to design this and the installation was completed in February 1888. The dynamo produced 70 A at 600 V and supplied power to the pump motor some 1,000 yards away. This, too, proved successful.

Since the original compressed air equipment was disconnected and not removed which became a commonplace practice later, to provide a standby facility, Snell had an ideal opportunity to evaluate both systems for efficiency. Two compressors, one horizontal and the other vertical, were available for testing. Efficiencies of 12% and approximately 14% were obtained. Snell conceded that the equipment was old; more modern and well maintained plant could increase these figures up to 20%. Even so, compressors could not hope to compare with the 44·4% efficiency of the electrical plant. These tests persuaded the mine management that the electrical plant was superior and another order was placed for an even larger installation to accommodate both pumping and haulage. The old compressors were finally removed and the foundations laid for two dynamos powering two motors.

The early electrical engineers in this field made admirable efforts to promote the wider use of electricity, but many were inconsistent over one very important point: cost. They either gave the costs of large installations or ignored them and treated the subject almost as an academic exercise. What the colliery owner wanted to know was, how much? D. Selby Bigge's costing figures for an electric plant to transmit 100 hp a distance of five miles were:

Generating plant: dynamo, exciter, foundation-rails,
belting, switchboard, instruments, regulating gear   £925
Line: copper cable, steel suspending wire, suspenders,
insulators, poles, lightning protectors, erection   £2,100
Receiving plant: motor, foundation-rails, switches,
regulating gear, instruments   £550

                                    Total cost   £3,575

This system had conductors of 19/12 SWG and a current of 103 A; the transmitting end voltage was 1,080 V and that at the receiving end 800 V. Selby Bigge also provided valuable information on losses, summarised in Table 3.2:

Table 3.2   Electrical efficiencies in transmitting 100 hp a distance of five miles

	hp	% bhp of engine
Brake horsepower of engine	168	–
Horsepower lost in belting	8	4.8
Horsepower lost in dynamo	11	6.5
Horsepower given out by dynamo	149	–
Horsepower lost in line	39	23.2
Horsepower received by motor	110	–
Horsepower lost in motor	10	6.0
Brake horsepower of motor	100	59.5

Not all collieries were large, however, and their requirements and available capital would not fall into this range. They received some guidance on running costs when one colliery, the Shirland Colliery of the Blackwell Colliery Company, fed back their findings in a short paper to the Institution of Mining Engineers in January 1892.[29]

Traditionally the small quantities of water, which collected in various places in the dip workings of the Shirland Colliery, were kept under control by pumping and barrelling by hand. The inconvenience and high cost of this led the mine management to install an electric generating plant for lighting and pumping.

The 24 hp compound-wound dynamo, driven through belt gearing from a high speed Robey engine, provided power to a 2 hp Davis and Stokes safety motor. This in turn, operated a Warner horizontal three-throw pump capable of lifting 2,000 gallons per hour against a 100 foot head. Both the motor and pump were made semi-portable by mounting on a pitchpine frame. The motor originally used was of normal construction, with the brushes on the outside of the commutator but, because of the dangers of sparking and arcing underground, it was replaced by a Davis and Stokes patent safety motor. This had its brushes inside the commutator, which was housed inside a brass end cover with a flame-tight joint. Such an arrangement confined the ignition, by sparking, of any explosive mixture present in the commutator. Further details of the construction are given in chapter 4.

Tests performed by the mine management to compare the results of electrical pumping with the original hand pumping and barrelling showed that the annual running cost of the former was vastly superior, despite the dynamo being overrated (producing only 17 A instead of its full load capacity of 150 A). Even with all the colliery lighting taken into consideration, the dynamo still only attained a 50% load.

The running costs of this installation were:

Value of fuel (90 tons) at 3s. per ton	£ 13 10 0
Stores, repairs, labour	£ 50 0 0
Interest and depreciation on proportion of outlay due to pumping plant at 10 per cent	£ 40 0 0
Total cost	£103 10 0
Previous cost of hand pumping, and barrelling same quantity of water	£657 0 0

3.5 Distribution systems

In 1890 Albion Snell reviewed the whole subject of electrical distribution in mines, together with the various methods used.[30] Snell, a vigourous campaigner for the use of electricity in mines, began on a positive note by saying that in view of:

> 'The results obtained in pumping and hauling at St. John's Colliery, Normanton, and in numerous other pits in Great Britain, Germany, France, and America, have at least been recognised as commercial successes. And it may now be fairly assumed that electricity is one of the recognised means of transmitting power in mines.'

He continued by saying that, with the exception of the dip pumps at Andrew's House Pit, no attempt had been made in Britain to install a parallel motor system underground. This was not because of any technical difficulty; electrical engineers simply opted for the easiest and cheapest system — a single motor and dynamo.

Despite the ' . . . cautious conservatism of . . . the British as a race . . . ' and the fact that this ' . . . elementary system of installation . . . ' had served well in the past, the time had come for a more general solution to electrical distribution.

Snell advocated three methods of distribution, viz:

(i)   Single transmission by series-wound dynamo and motor
(ii)   Constant current transmission, applicable usually to one dynamo and a number of motors, all running in series with each other
(iii) Constant potential transmission, comprising in its simplest form one shunt or compound-wound dynamo and a number of series, shunt or compound-wound motors connected in parallel.

The parallel system, in Snell's opinion, was best suited for general distribution in collieries. It had one major disadvantage: the cost of copper mains when a relatively low voltage supply was used to run both large and small drives. Under such circumstances he advocated the use of direct current transformers (dc converters). Such equipment, although resembling a motor with two armatures, was effectively a dynamo–motor combination, and had never been used underground before. Snell could see no objection to their use, provided that they were locked in cabins in the main intakes and inspected regularly. The cost of these transformers was estimated at £8 per horsepower, but they would still be cost-effective when the saving of copper mains cables and the reduced size of the motors were taken into account. As regards distribution voltage levels, he had previously used 725 V (albeit with a separate pair of machines), but thought it most unlikely that anyone would want to exceed 500 V on an extended distribution system. That voltage would be higher than was needed for motors powering rock drills, coal cutters, dip pumps and auxiliary fans. Ideally, from both the economic and safety view, the large motors near the pit bottom should have a high potential, with lower voltages for the smaller motors, fed from dc transformers and located further out into the workings.

Discussing the cable for such installations, Snell recognised the temptation of using uninsulated returns since there was always plenty of old cable and rope lying around. He admitted to doing this in the past, even though the risk of short circuits would increase whenever a motor was added to the system. 'Considering all things', he advised against the use of naked returns 'where more than one or two motors are run off the same dynamo, particularly if the tension be high or the power large.'

A constant potential system was installed in 1890[31] at the Crawshay Bros. Newbridge Rhondda Colliery, in South Wales. Snell personally superintended the installation and boldy described it as one of the more important plants running in Great Britain. He could see 'no reason to prevent it being equally successful under similar conditions in other pits'. This claim was based on the use of an over-compounded dynamo to maintain as near a constant system potential as possible, so that either of the two 7 bhp pump motors could be switched on and off without significantly affecting the other. The 'compound' field winding, patented jointly by R.E.B. Crompton and Gisbert Kapp in 1882, enabled a generator to provide a constant output voltage to varying loads without continually adjusting the brush position, previously necessary whenever a change in load occurred to reduce sparking at the commutator.[32] With the over-compounded dynamo, a rheostat was placed in the shunt field and was adjusted to 'fix' a particular voltage. Doubts were raised as to the reliability of such a method and, more specifically, the ability of the workmen at the Newbridge Rhondda Colliery to perform such a task. Snell had no doubts whatsoever on this score and told his critics, it 'was thoroughly understood at that colliery'.[33]

The same system also provided electric lighting to the pumping stations, pit bank and pit bottom. It was an excellent example of a mixed purpose distribution system employing some of the most up-to-date technology. It certainly impressed the management of the Newbridge Rhondda, for they hoped to extend the use of electricity at an early date. Albion T. Snell, as well as being a thoroughly competent engineer, was also no mean salesman!

A variation of this system was installed soon afterwards at the Earl of Durham's collieries at Bunker's Hill.[34] An installation described by D. Selby

Bigge as:

> 'probably the largest and most complete in this country . . . used for performing several of the most important operations necessary in colliery work.'

There was no doubt that this was an important installation, as the following description reveals. Since, however, such claims appear with unerring regularity by engineers subsequently describing their work, they must be treated with a certain amount of circumspection.

The Bunker's Hill installation comprised two Willans high-speed engines (each developing 140 hp at a speed of 380 rpm) connected through clutches to a countershaft. From this shaft, via pulleys and belts, the dynamos were driven. The two power dynamos, of Goolden manufacture with Gramme wound armatures, produced 80 A at 780 V when rotating at 500 rpm. They were excited by a smaller dynamo, which gave 90 A at 100 V when rotating at 1,200 rpm. Lighting was also provided by this dynamo. A fourth dynamo (giving 120 A at the same voltage) acted as a reserve to the latter.

The two power dynamos were connected in parallel, and the rheostats adjusting the field (magnetising) currents in order to match their output voltages were mounted on the main switchboards. The outputs from these dynamos were fed into the switchboards and monitored by ammeters and voltmeters, before passing through the main insulated double-pole switches (of the lever type), the automatic switch and down the shaft on a cable (equivalent in area to a 19/14) to the distribution point.

From cast-iron junction boxes, having gas-tight covers and internal terminal clamps, the circuit was split into two. Each was isolated by knife (or lever) switches which could be opened (or so it was claimed) without the slightest sparking. One circuit using 7/16 equivalent cable fed a pump 670 yards away. The other branch, in 19/16 equivalent cable, ran 1,290 yards to a haulage engine, and then another 1,500 yards to a winding engine. The cables throughout were of the Atkinson patent safety pattern and insulated with two layers of india-rubber tape, wrapped in opposite directions. Since the pit was perfectly dry, it was considered unnecessary to go to the expense of using vulcanised rubber and compounded hemp braiding insulated cable. The cables were carried through the workings on porcelain insulators, affixed to the roof or timbers, with one run on either side of the road. Although this is not stated, the cable would almost certainly have been secured by rope or leather thongs which were designed to break in the event of a roof fall. Since it was hazardous to make soldered joints underground, the lengths of cable were connected by terminals fastened onto a slate base and enclosed in cast-iron boxes.

The pump, manufactured by Goolden, was a standard mining pattern three-throw type, capable of lifting 150 gallons per minute 270 feet vertifical lift through 6-inch pipes. Fig. 3.2 is a typical example. Fastened to this, through gearing, was a 24 hp motor, controlled by a resistance starter housed in a strong cast-iron, gas-tight case.

The haulage was of the continuous type – again of Goolden manufacture — fitted with a patent enclosed type motor, capable of developing 40 bhp at 650 rpm, to drive a Hurd patent clip pulley five feet in diameter through two sets of cast iron gearing with double helical teeth. The clip pulley was used in

Fig. 3.2   Goolden safety motor driving three-throw dip pump
Illustration taken from *Electric motive power* by A.T. Snell, p.333 (shown in both 1894 and 1899 editions)

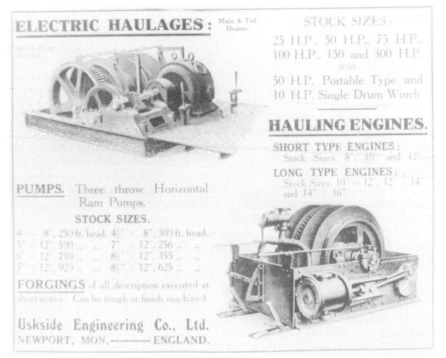

Fig. 3.3   Advertisement c. 1921 detailing haulages manufactured by the Uskside Engineering Co., Ltd., Newport

preference to a coned drum to reduce the wear on the rope which, at full speed, was capable of travelling at four miles per hour.

Whilst electric pumping and haulage was not an uncommon feature in colliery installations at this time, electric winding engines certainly were. This must have been one of the earliest examples working in a British coal mine.

The winding engine resembled the haulage gear, except that the clip pulley was replaced by a 4 feet diameter drum having a central dividing flange. Two hand levers, mounted on a wooden platform behind the machine, controlled the winder: starting, stopping, braking, reversing and regulating the speed. The engine was designed to wind in a staple from a depth of 144 feet: each lift weighed some 10 cwt. The Goolden motor was rated at 32 bhp; but this was considerably over-rated for the loads raised.

Ancillary electrics at the colliery comprised lighting (about fifty 16 cp incandescent lamps located up to 400 yards in-bye) and a combined telephonic and bell-signalling system, which ensured communication throughout the whole pit, including the underground electric stations and engine-bank.

3.6 dc systems

Industry, and the mining industry in particular, never concerned itself as much as public supply companies over the choice of alternating or direct current. Most industries adopted dc systems from the outset. Even the introduction of the three-phase turbo-alternator at the beginning of this century failed to persuade many colliery owners to choose ac.

The reason why direct current had, by the early 1890s, established such a dominance in power distribution systems was largely historical, from Gramme's first practical machine and the way Gramme and Fontaine demonstrated in Vienna that such machines could be used together in a reversible combination of motor/generator mode.[35] It was only natural that subsequent development work by Gramme and others should use the same system. The unavailability of a practical ac motor until after 1888 considerably strengthened their choice.

After 1880 dynamos were developed rapidly. Efficiencies of up to 95% were soon being obtained. This, together with improved mechanical construction enabling them to run for long periods without overheating or failing, made them a very attractive proposition. Outputs also increased considerably. In 1885 a vertically-driven machine could deliver 100 kW; within five years, a machine of 1½ tons capable of producing 1,500 A was on the market.[36] DC had a number of attractive features: wide speed control, the paralleling of generators onto the same busbars, battery back-up when the generators stopped running (rarely adopted by collieries unless they had electric locomotives), choice of working voltages by means of two, three, four or even five wire supplies, and the availability of reliable, low speed prime movers. It is easy to get the impression that dc machines had few problems. Although a large number were reliable and gave excellent service, they were certainly not infallible, as the records of the National Boiler and General Insurance Company Limited indicate.[37] An analysis was taken between 1898 and 1902 of machine reliability covering several thousand machines ranging in size from 0·5 hp motors through to 800 kW dynamos, under what the author, an engineer of the Insurance Company described as, 'somewhat better than average conditions, so far as superintendence is concerned'. The survey was not confined to colliery installations but extended to industry in general, and Table 3.3 summarises the findings.

Table 3.3 Percentage breakdown of dynamos and motors in industry between 1898 and 1902

Points of breakdown	Percentage
Mechanical	11·03
Brush-gear, leads and terminals	8·6
Field-magnets	8·58
Armatures	47·8
Commutators	14·9
Starters	4·7
Unkown	3·13
Various	1·56

Although armature problems accounted for almost half of the breakdowns, the author considered that overloading was responsible in very few cases (1·37%). The two biggest problem areas were: construction (39·36%) and maintenance (31·46%). Of the former, more than 18% was attributable to bad design and almost 14% could be put down to bad workmanship. On the basis of such figures one could well understand the reluctance of colliery owners to install motors underground. All of the blame must not be laid on the manufacturers, but it is apparent that they had little awareness of the special considerations applying to mining engineering. Bad maintenance accounted for more than 31% of subsequent breakdowns. Of this figure, 22·5% was blamed upon 'defective attention'. Although much information could be extrapolated from this insurance company's figures, the most pertinent comment came from Robert Nelson, who, in 1939, when reviewing the state of electricity in mines thirty years earlier, said:

'Electrical and mining engineers had still to educate themselves and each other: if the former had not yet grasped the differences that exist between surface and underground conditions, the latter, sound enough as mechanical engineers, had no knowledge of electricity . . .'[38]

There was no mistaking the fact that he was addressing the Institution of Electrical Engineers, because in the same sentence he said:

'. . . and some amongst them harboured the delusion that bell-hanging methods could usefully be imported below ground.'

As the load on distribution systems increased, one disadvantage of dc became very apparent — voltage drop. There were a number of remedies:

• Compensate for the losses by installing additional dc generators in the system

- Generate at high voltage dc and reduce the potential locally by rotary transformers
- Generate at high voltage ac and reduce the potential locally into dc by rotary machines
- Change from dc to ac and distribute at as high a voltage as possible to reduce losses. Local transformers would reduce the potential to the required working level.

The majority of colliery owners opted for increasing the number and/or size of dc generators. This was not an unnatural response since it was less complicated. It was not necessarily the most cost effective.

The real answer was, of course, ac transmission and distribution. But, for power purposes, such technology was still in its infancy.

3.7 ac systems

Within two years of the successful demonstration of high voltage dc transmission at the Munich Exhibition, an equally successful demonstration took place at the 1884 Turin Exhibition in Italy, this time with ac.[39] Credit for this went to Lucien Gaulard, a Frenchman, at the time working on the lighting of London's underground system.

A 30 kW, 2,000 V, 133 Hz generator, driven by a steam engine, provided power to a loop 80 km in length, containing three series-connected, single-phase transformers. Their secondaries supplied various railway stations.

The installation was tested by a panel headed by Professor Galileo Ferraris, of Torino University, and the efficiency was found to be about 90%, which proved that transmission of ac power over long distances, withstanding moderate losses, was possible. For his efforts, Gaulard was awarded a prize of 15,000 lire.

Another exhibition, at Frankfurt am Main in 1891, saw the world's first three-phase power transmission and distribution system successfully demonstrated.[40] A 210 kW, 55 V, 40 Hz, three-phase generator, built by the Swiss company of Maschinenfabrik Oerlikon, supplied power through transformers which stepped the voltage up to 15 kV. At the receiving end of the 170 km overhead line, in the exhibition hall, the voltage was transformed down to 65 V (phase-neutral) to supply one thousand 16 cp incandescent lamps.

A number of three-phase motors, of AEG manufacture, were also connected to the system. The largest (100 hp) drove a pump for the waterfall and the smallest (0·1 hp) a fan. On the last day of the exhibition a 20 hp synchronous motor was installed.

The outstanding success of this demonstration, together with the absence of a commutator and hence no sparking on polyphase motors, should have had an immediate appeal — particularly to the mining industry. In reality, no great changes occurred because of the established position of dc and other problems peculiar to ac. Unlike dc generators, which run readily in parallel, early ac alternators did not and the following (well quoted) description of two alternators not settling down until they had 'jumped for three or four minutes' after being connected in parallel, well illustrates the problem.[41] Such jumping would have done little for the life of the lamps connected on the system and probably even less for the constitution of the operator.

The principles of synchronising were described as early as 1868 by Wilde and developed by John Hopkinson in 1884. Now two ac alternators could run perfectly satisfactorily in parallel and share the load, provided that they were carefully synchronised first and their ouput waveforms were compatible. However, when output waveforms of these machines were liable to vary,[42] then successful synchronisation became very much a question of luck.

Until the outputs of alternators and the quality of prime movers improved, the practice of running single alternators persisted.

3.8 Drives and driving power

3.8.1 Belt and rope drives

As the loads increased greater demands were placed upon the dynamos. Increased output could be obtained from these in a number of ways: by increasing their speed, or installing larger or multipolar machines. Since these early systems invariably used bipolar machines and installing new equipment was expensive, the obvious choice was to increase the speed (provided that the capability of the machine was not exceeded). Belt and rope drives were one answer, but whilst flexible they had limitations, not least transmission losses through the belts etc. and taking up valuable floor space. The belt drive of a 50 hp motor used for pumping and haulage at the Dumbreck Coliery, Kilsyth, in 1893, for example, was operating with 23 foot centres,[43] Many collieries, by virtue of their geographic location, never had such space. A similarly rated direct-coupled dynamo, manufactured by Messrs. John Davies & Son, which occupied an area of 8 ft × 4 ft,[44] was very attractive on this consideration alone.

Direct-coupled high speed engines (typically Willans), both of the horizontal and vertical design, driving bipolar dynamos at 300–500 rpm, became a very popular and reliable means of generation over the next few decades, provided their limitations were recognised. They could not cope with widely varying loads, as required in a small winder; and needed a steam pressure above 80 psi.[45] They were best for lighting applications where the load could be predicted so they could be arranged to be run at or near full load. Increases in load from other sources could be catered for by bringing additional machines on line.

As the demand for electricity increased still further, the main disadvantage of the high speed bipolar dynamo became patently obvious. It was simply not feasible to build machines of high voltage and capacity in a bipolar design. They would need such large armatures that these could not be safely rotated at the speeds needed to produce the electromotive forces required. The manufacturers therefore developed large, efficient, slow speed engines, with multiple cylinders directly coupled to multipolar dynamos. These would obtain maximum benefit from the steam. Such a dynamo offered additional benefits. The more effective distribution of flux made them economical in construction materials. These large units attracted the attention of the electricity supply industry, but again only large collieries, contemplating central generation, opted for them. The remainder were content to obtain their electricity from the high speed machines.

The form of drive was also very important with alternating current, because a slow speed reciprocating engine needed a multipole generator to produce an alternating output at an acceptable frequency since

$$\text{speed of rotation} = \frac{120 \times \text{frequency}}{\text{number of poles}} \quad \text{rev/min}$$

Thus, in order to obtain an output of 50 Hz, a 12-pole generator would have been driven by a reciprocating engine at a speed of 500 rpm. This was, effectively, the maximum speed of such a machine. Belt and rope drives could increase the speed but, as with dc, these involved more space, additional costs and losses in efficiency.

For the full benefits of ac to be realised, a high speed prime mover was required: that was to come in the form of the steam turbine.

3.8.2 The steam turbine

One writer, in 1962, described the invention of the steam turbine, with its directly connected generator, as:

> 'the most potent factor in the history of electrical engineering throughout the world over the past sixty years.'[46]

This was no exaggerated claim, for the steam turbine revolutionised the electrical industry. Power could suddenly be generated in hitherto undreamed of quantities.

Charles Algernon Parsons, the sixth and youngest son of the Earl of Rosse, in about 1884 successfully produced a steam turbine, which consumed steam of about 150 lb/kWh, and enabled a dc dynamo to produce 75 A at 100 V when rotating at 18,000 rpm.[47] This machine enjoyed a long and reliable life at a lampworks in Gateshead[48] and is now preserved at the Science Museum, South Kensington. Encouraged by this early success, Parsons went on to develop larger and more efficient turbines. By 1890 the Newcastle and District Electric Light Company Ltd had commissioned two 75 kW, 80 Hz, single-phase, 1,000 V turbo-alternators which ran at 4,800 rpm.[49] Other machines, each progressively larger, were installed in power stations both in Britain and abroad. Despite their obvious advantages (simplicity, robustness and economy with steam under varying load conditions), they met a certain amount of opposition, like most innovatory equipment. However, an authoritative report, published in 1892 by Professor Ewing of the University of Cambridge, which extolled the virtues of alternators, did much to dispel imagined fears and ill-based prejudices, and so helped to establish the turbo-alternator as the new standard for the large scale generation of electricity.[50]

Ironically, it was not a power station that claimed the distinction of installing the world's first three-phase turbo-alternator, but a colliery — the Ackton Hall Colliery in Featherstone, near Leeds, in 1900.[51] The proprietor of the colliery, Lord Masham, took an active interest in the development of electrically operated coal-cutting machines. He saw the turbo-generator, which developed 150 kW at 2,520 rpm and supplied current at 350 V between phases at a frequency of 42 Hz, as the ideal solution to providing the necessary power. Roslyn Holiday was the colliery's electrical engineer.

Ackton Hall was no newcomer to turbo-generators. As early as 1895 they had installed two such machines, albeit of the dc type, each developing 150 kW at 500 V. In 1899 they demonstrated a further commitment to turbo-generators by putting in a third set. In 1901 yet another dc type, having an ouput of 300 kW, was installed. By 1918 this colliery could boast no fewer than eleven Parsons turbo-generators, demonstrating that collieries could adapt the most modern technology for their benefit. Ackton Hall was, however, the exception rather than the rule.

3.8.3 Driving power

In 1882 Sydney Ferris Walker made some self-evident statements when he defined the Model Electric Light Engine as one which, once set to a particular speed, *never* varies.[52] He continued to explain that any engine, provided that it was strongly constructed, would do the job, if the following criteria were satisfied:

- That a good supply of steam was available
- That the flywheel was fairly heavy
- That the governor always cut off the steam at the proper pressure

For large electrical installations in collieries (referring only to lighting), he saw a fan engine as the ideal answer. This was a classic case of a feature common in colliery work — 'spin-off' benefits arising out of the use of improved techniques. In this instance, colliery owners were already installing steam-driven fans of ever increasing capacity to provide more air to enable more expansive working or access to deeper seams. Smaller installations could be catered for by any reputable manufacturer's steam engine, provided that it was fitted with a special high speed governor or an ordinary governor driven at high speed. He cited the installation at Harris's Navigation Colliery, Glamorgan, where a Kitson high speed engine was directly coupled to a generator. This had been working for twelve months and, in the electrical engineer's view, 'nothing could be better'. The electrical engineer at another South Wales colliery, Cymmer, no doubt wished that he could have expressed the same sentiment, but unfortunately he experienced large variations in the output of his incandescent lamps every time the winding engine drew a load, although the prime-mover was fitted with a heavy flywheel and governor.

Shortage of exhaust steam was not normally a problem; indeed, many colliery owners lamented the fact that energy was wasted in venting off to atmosphere. If, however, the boilers were old and inefficient and loaded to capacity with ancillary equipment, such as fans, compressors and workshop equipment, there might not be enough steam. Then, in order to secure a reliable source for the prime mover, colliery owners would install a semi portable engine with cylinders situated under the front end of a locomotive boiler (typically a Fowler).[53] This had the advantage of being mobile, compact and comparatively inexpensive.

The exhaust steam from large winders was normally sufficient to power relatively large electrical installations. When the quantity of exhaust steam *just* equalled the amount required to drive the prime mover(s) *and* all the ancillary equipment on the colliery site, a good case for electrification could usually be made (see chapter 8).

Permanent boiler installations using Babock & Wilcox or Lancashire boilers, which were popular in power stations, demanded large capital expenditure, so only large collieries, contemplating central generating facilities, were in a position to adopt them (again, see chapter 8).

From about 1880 another form of fuel became available to generate electricity:

gas. The future of gas engines seemed assured when, in 1881, Sir Frederick Bramwell told a meeting of the British Association at York that within fifty years steam machinery would only be found in museums; its place would be taken by the gas engine.[54] In reality this never happened because of the many problems associated with early gas installations, such as noise, vibration, difficulty in parallel running and inability to compete economically with steam plant which used cheap bituminous coal.[55]

A variation in gas generation was the use of waste heat and gas either through boilers for raising steam or directly in gas engines. This was more successful, but was only attractive in areas where large quantities of coke and coal were produced. Apart from a few notable exceptions in South Wales, this meant the North-East of England.[56]

3.9 Notes and references

1 KEITH, N.S.: 'Can we transmit power in large amounts by electricity?' *Transactions of the American Institute of Mining Engineers*, 1877–78, Vol. 6, p. 452
2 *ibid*; 1882, Vol. 10, p. 309
3 WALKER, S.F.: 'The principles of electric lighting and transmission of power by electricity: Section D — Transmission and distribution of power'. *Transactions of the South Wales Institute of Engineers*, 1882–83, Vol. 13, p. 283
4 TOMOS, D.: '*The South Wales story of the Association of Mining Electrical and Mechanical Engineers*', Cardiff, 1960, p. 15
5 POULTER, J.D.: '*An early history of electricity supply. The story of the electric light in Victorian Leeds*', Peter Peregrinus, Stevenage, 1986, p. 21
6 WALKER, S.F.: *op. cit.*, 'Section B — Dynamo electric machines', p. 163
7 PARSONS, R.H.: '*The early days of the power station industry*', Cambridge, 1940, pp. 3–4
8 BOWERS, B.: '*A History of Electric Light & Power*', Peter Peregrinus, Stevenage, 1982, p. 90
9 JARVIS, C. MacK.: 'The generation of electricity', in '*A history of technology*', Ed. Charles Singer *et al.* Vol. V, Oxford, 1958, p. 192
10 BOWERS, B.: *op. cit.*,pp. 90–91, 248–249
11 The abstract appeared in the *Proceedings of the Institution of Civil Engineers*, 1882–83, Vol 71, p. 509 under 'Foreign Abstracts'. It was taken from the *Bulletin de la Societe de l'Industrie Minerale*, 1882, Vol. 11, p. 5. The report in the *Electrical Review* appeared in 1882, Vol.11, p. 149, where the mine was referred to as La Perroniere and the gradient of the incline given as 1 : 2·5 and not 1 : 25 as quoted by Keith.
12 Chatterton's compound is an adhesive insulating substance consisting largely of gutta-percha
13 The exchange rate at that time was 25 francs to £1
14 *Electrical Review*, 1882, Vol. 11, p. 150
15 *Proc ICE*, 1882–83, Vol. 71, p. 513
16 DUNSHEATH, P.: '*A history of electrical engineering*', 1962, p. 159
17 *Minutes of Evidence taken before the Department Committee on the Use of Electricity in Mines*, 1904, Q5002, p. 143
18 *Trans. SWIE*, 1882–83, Vol. 13, p. 277
19 *Engineering*, 1883, Vol. 35, p. 22
20 *Trans, SWIE, op. cit.*, p. 281. By comparison with the cable at La Perroniere, this cable cost £100 per 1,000 yards.
21 TUCKER, D.G.: 'Early electrical systems in collieries: The Trafalgar Colliery in the Forest of Dean and the Brain Family'. *Proceedings in the 1975 IEE weekend meeting on the history of engineering*, 1975, p. 4
22 *ibid.*, pp. 4–5
23 *Proceedings of the National Association of Colliery Managers*, 1889, Vol. 1, p. 271: *Engineering*, 1887, Vol. 44, pp. 603, 625, 685: *Engineering*, 1888, Vol. 45, pp. 9, 73: *Trans ICE*, Vol. 104, Part 2, pp. 121, 133: *Transactions of the Midland Institute of Mining, Civil and Mechanical Engineers*, 1887–89, Vol. 11, pp. 342, 344, 346

24 *Electrical Engineer*, 1889, Vol. 3, p. 510
25 *Electrical Review*, Vol. 25, p. 558
26 *Electrical Engineer*, 1888, Vol. 1, p. 298
27 *Proc. NACM*, 1889, Vol. 1, p. 195
28 BIGGE, D.S.: 'The practical transmission of power by means of electricity, and its application to mining operations', *Transactions of the Institution of Mining Engineers*, 1891–92, Vol. 3, p. 288
29 DEACON, M.: 'Notes upon a small electric pumping plant', *Trans. IME*, 1891–92, Vol. 3, pp. 191–95
30 SNELL, A.T.: 'The distribution of electrical energy over extended areas in mines'. *Trans. IME*, 1889–90, Vol. 1, pp. 142–56
   Albion T. Snell (1858–1936): After a short period of pupilage under James (later Sir) Swinburne, Snell joined the staff of Hammond & Co. In 1885 he was appointed Chief Engineer to Messrs. Immisch & Co. With this company, his work was mainly concerned with the design of motors and dynamos, and their application to mines. When, in 1890 the business of Messrs. Immisch & Co. was absorbed by the General Electric Power & Traction Co., he became Chief Engineer of the new company and turned his attention to battery-driven tramcars. He later set up his own firm of Albion T. Snell and Partners.
31 SNELL, A.T.: 'Notes on electrical work in mines',. *Trans. SWIE*, 1890–91, Vol. 17, pp. 196–198
32 BOWERS, B.: *op. cit.*, p. 93
33 SNELL, A.T.: *op. cit.*, p. 239
34 BIGGE, D.S.: *op. cit.*, pp. 291–295
35 BOWERS, B.: *op. cit.*, p. 247
36 DUNSHEATH, P.: *op. cit.*, p. 113
37 CORMACK, A.C · 'Electrical plant-failures: Their origin and prevention', *Trans. IME*, 1902–03, Vol. 25, pp. 548–549, 571
38 NELSON, R.: 'Electricity in coal mines: A retrospect and a forecast', *Journal of the Institution of Electrical Engineers*, 1939, Vol. 84, p. 598
39 'Crosstalk', *Electronics and Power*, Aug. 1985, p. 551
40 *ibid.*, pp. 551–552
41 DUNSHEATH, P.: 1962, p. 169
42 *ibid.*, p. 170
43 FORGIE, J.T.: 'The electric power plant at Dumbreck Colliery, Kilsyth', *Trans. IME*, 1893–94, Vol. 7, Scaled off sketch plans
44 GRAVE, L.W. De: 'Continuous current dynamos and motors', *Trans. IME*, 1894–95, Vol. 9, p. 197
45 *ibid.*
46 DUNSHEATH, P.: *op. cit.*, p. 196
47 PARSONS, R.H.: *op. cit.*, p. 170
48 DUNSHEATH, P.: *op. cit.*. p. 197
49 PARSONS, R.H.: *op. cit.*, pp. 171–172
50 *ibid.*, pp. 174–175
51 PARSONS, R.H.: *The development of the Parsons steam turbine*, 1936, pp. 57–59
52 WALKER, S.F.: *op. cit.*, 'Section B — Dynamo electric machines', pp. 188–189
53 PARSONS, R.H.: *op. cit.*, 1940, p. 17
54 *ibid.*, p. 151
55 *ibid.*, pp. 160–161
56 The Powell Duffryn Steam Coal Company, in South Wales, installed a gas engine, in 1896, at their Bargoed Colliery. This engine operated from the waste heat and waste gas produced from their re-generative coke ovens and by-products plant. *The Powell Duffryn Steam Coal Company, Limited, 1864–1914*, p. 23
   In evidence given before the Royal Commission on Coal Supplies, it was estimated that if gas engines were used exclusively for power purposes, then, after the requirements of the stoves and blowing engines had been met, there would still be available from the Cleveland furnaces a continuous supply of surplus gas equivalent to 61,000 hp. Quoted from MERZ, C.H.: 'Power supply and its effect on the industries of the North-East Coast', *The Electrician* 2 October 1908, p. 961

Chapter 4

Recognition of the hazards of electricity

4.1 Outlining the problem

We have seen in chapters 1 and 2 that, over the years, firedamp had proved a very serious hazard to the miner, particularly where naked lights were used and even where safety lamps had been employed. It has also been shown that electricity, both mains supply and in the form of the secondary cell powered safety lamp, offered the hope of eventually solving this traditional lighting problem.

Electricity was also seen by some as an ideal energy source to supply the increasing demand for underground mechanisation. The independent discovery of the principle of the self-excited generator in 1866 by Wilde, Varley, Siemens and Wheatstone, and successfully developed by Gramme, Siemens and others, offered an efficient and convenient method of generating electricity for a broad range of industrial applications.

As shown in chapter 3, both the ease of transmission over relatively long distances and its high transmission efficiency were seen as attractive properties of this new motive power, and a few mining engineers recognised its unique possibilities.

These advantages were, however, accompanied by various potentially hazardous features. In the case of mining, these were principally associated with the ignition of firedamp. The main source of danger was the sparking commutator, with cable or equipment faults providing secondary, though still serious risks. Many mining engineers and managers were therefore sceptical about using electricity, particularly in fiery mines. Fears about electricity were usually voiced by engineers, often with long years of experience in mining, who might not claim a deep understanding of this new branch of technology, but were thoroughly practical men and recognised potential sources of danger in the mining environment. Their warnings would have been heard throughout the industry.

One such was J.B. Simpson who, in his presidential address to the North of England Institute of Mining and Mechanical Engineers in 1891, said[1] that mining engineers would be very sceptical until 'absolute experiment has satisfied them that the desired-for result has been obtained.'

In at least one pit, workmen used the static electricity generated by the belt drives of machines to ignite gas at lighting fittings. This was done by holding the finger of one hand close to a revolving drive belt, while in the other hand a metal rule was held near the metal gas fitting. The arc thus discharged lit the gas.[2] If this type of electric spark could ignite town gas, engineers reasoned that sparks from a motor commutator could ignite pit gas.

The Assistant HM Inspector for Mines for the Yorkshire and Lincolnshire District, J. Gerrard, expressed the feelings of many of his colleagues when, at a meeting of the Midland Institute of Mining Civil & Mechanical Engineers in January 1889, he said that, although electricity was still in its infancy, great advantage had already been gained from it, both in signalling where:

'in fact it would be impossible to get through the work with the old form of signalling on our modern engine-planes'

and in lighting:

'upon the pit tops beautiful installations of electric light . . . and the flaming lamps and open torches which lit the bottom of our shafts and threw that lurid, flickering light only a very short distance away from the hanger-on and his helpers, has now passed away.'

Having seen what electricity can do, he considered that:

'we must be disposed to give every help and countenance to this further development of electricity, the using it as motive power . . . We are taking out open lights, we have made vast improvements with regard to safely lighting our mines, we are taking out furnaces, going in for mechanical ventilators and so on, and we must not introduce electricity as a motive power with a danger and liability to sparks, if there is a danger attached to it, without that being brought to the minds of those inventors that they may grapple with it and overcome it.'[3]

He thought that compressed air, whilst being 'very comfortable and cool' to use was a very expensive power, and though a good case for the use of electricity had been made on economic grounds, it had not been proven to be safe. In concluding his address he suggested that 'conclusive tests with regard to sparks' should be carried out, that the dangers be acknowledged and hopefully overcome.

Gerrard returned to the next meeting of the Institute in February 1889 and stated that he had heard of tests being carried out with two types of batteries, one a Leclanché type with twelve cells and the other a Bunsen. With both batteries, sparks were produced by short-circuiting a wire about five yards long. With the Leclanché battery, a spark of about one-twelfth of an inch long was produced which did not ignite an explosive mixture of coal gas and air, whereas the Bunsen battery's spark was one-eighth of an inch long and fired the gas. He agreed with A.T. Snell that under normal conditions, with machines working as they ought, the commutator would produce only low temperature sparks but if something were to go wrong, the intensity of the sparks would increase to produce a dangerous situation:

'We must not shut our eyes to the fact that plant may get out of order . . . On ordinary occasions, the sparks may be harmless but if something goes wrong, if there is an irregularity, you may have a dangerous flame. This should be thoroughly elucidated by experiments or provided for and made impossible by further appliances.'[4]

Such serious comments from Gerrard, the District Inspector, would have been heeded by all concerned, whether designers or potential users. In the Midland area at least, the use of electrical machines in-bye in fiery mines would not be acceptable without special measures to combat the hazard of sparking commutators.

Gerrard's statement was made during discussion of two papers read before the Institute on 21 December 1888. The first was by G. Blake Walker on 'Electricity as a Motive Power' and the second by A.T. Snell on 'An Electric Locomotive for Mines'.[5] G.B. Walker, engineer at the Wharncliffe Silkstone Colliery, Barnsley, and Vice President of the Institute, replied that:

> 'with regard to this scare, if I may call it . . . I think we may dismiss without much anxiety this question of sparks, if you have for instance, an intake airway where 30,000 cu ft of air per minute is passing, even if it be a mile or more from the pit bottom, I think myself that to raise a great outcry against the making of sparks in such conditions as that is excessive caution.'

The depth of feeling and diversity of opinion on this question can be gauged from the comments of C.E. Rhodes of Altwark Main and Carr House Collieries and President of the Institute, when he retorted:

> 'One remark Mr Walker made in the beginning of his address I personally take exception to and that is that this question of sparks should be called a scare.'

In his opinion, it did not matter where in the pit electrical apparatus was put, because he had seen gas backing up against more than 30,000 feet of air, fouling fresh air to within 650 yards of the pit bottom, concluded:

> 'You cannot get over facts of that sort . . . I think the subject should be thrashed out, if the sparks are dangerous try to remove them; if we cannot remove them, it is a strong point against electric haulage.'[6]

How extensive was the problem of gas in British pits at that time? Throughout the coal field, the amount of gas encountered at different pits varied quite markedly. In some pits, firedamp was rarely, if ever, found, whilst at others gas was constantly seeping into the workings. In some mines, men could be met by a sudden onrush of gas bursting from the seam, sometimes with sufficient force to bring down large sections of the working face. The pressure behind these jets of gas, or blowers, varied considerably. At a meeting of the Institution of Civil Engineers in November 1887, Col. J.D. Shakespear told members that at one pit on Tyneside the pressure of a blower had been recorded at 600 lb/sq in, at another 461 lb/sq in. Compared with these, the wind pressure in a hurricane of some 50 lb/sq ft was 'a trifle'.[7]

With regard to quantity and consistency of supply, a more extreme example was that of C Pit at Wallsend (Fig. 4.1). Here, from 1834 to 1853 a section of the pit extending over some 50 acres had been walled up and the space ventilated to the surface through a 4 in diameter pipe.[8] Over this period, the gas 'flared off',

Fig. 4.1 Wallsend colliery
(Source: Coal mining history reprint by David Books 1987. *Views of the Collieries in the Counties of Northumberland and Durham* by T.H. Hair, first published 1844)

had maintained an average flow rate of 66 cubic feet per minute, or some 34·5 million cubic feet per year. By 1853 the flow rate had reduced to 34 cubic feet per minute and gas was still encountered in the still productive areas of the pit.

These may be extreme examples of conditions in gassy pits, but it is little wonder that, where the menace of firedamp was present, mining engineers and managers looked on the introduction of electricity with great caution and scepticism.

4.2 Initial attempts at solutions

As shown in Table 2.1, by 1889 out of some 4,309 pits in Britain, electricity had been applied to lighting at only 42 and to power drives in about 6. Collective experience indicated that electricity had a considerable advantage over other power sources in efficiency of transmission. It could not be rivalled as a means of providing light of various intensities and easily controlled, but its future in coal mining would be seriously limited if the dangers associated with the possible ignition of gas could not be contained.

With only a handful of installations in Britain, engineers with broad experience of the various aspects of mining electrical installations (motor design, control, pumping, haulage, coal cutting, cable design and installation, etc.) were difficult to come by. There was no organisation to identify problems or carry out research and development work. If these tasks were to be tackled at all, they would have to be undertaken by either or both the equipment manufacturers and the users.

From contemporary accounts, it is clear that in Britain at this time the Immisch Co. of London was taking a leading role. From the comments of their chief electrical engineer, A.T. Snell, they were committed to the future development of electrical applications in mining and, therefore, to safe plant and equipment.

Snell was questioned about the possible dangers from sparking commutators following his paper to the MIMC&ME in December 1888. He said that development work was taking place and improvements were being made, but from the information he presented, it appears that such progess was not following any clearly defined plan based upon a proper understanding.

The proceedings of the MIMC&ME show that from December 1888 to the middle of 1889 the use of electricity in mining was a topic of considerable interest. The advantages of electric power may have been demonstrated, but its inherent dangers were also recognised, though not always admitted by those advocating its adoption. Some of the solutions suggested for the danger from sparking electrical plant were barely credible. A.T. Snell evidently felt under pressure to reply to the various points raised and judged a poor response was better than no response at all.

A new pumping plant of about 20 hp, he said,[9] had been installed in South Wales (at the Llanerch pit), using a new type of machine which was an improvement on those supplied to Locke & Co. at Normanton (the St John's pit) between 1887–88. The sparking of the commutator having been so greatly reduced 'that it is difficult to see the sparks even in the dark'. Exactly how this improvement had been achieved was not explained at this meeting.[10] Earlier, in what appears to have been a half-hearted attempt to play down the problem of arcing between brushes and commutator, Snell commented:

> 'when we take a dynamo or motor, the sparks are in contact with such a large body of metal that the temperature is low . . . I do not mean to say it would not ignite gas under certain conditions but, I mean to say this — that although you see a few sparks, it does not follow there is danger'[11]

When reminded that not so long before he contemplated putting a motor in an explosion-proof box, he thought perhaps there had been some misunderstanding; if he had considered this idea, 'he had certainly grown wiser since then'.[12]

His intention was now to house the motor in a box that was almost airtight, with a one or two inch pipe connected to an opening at the bottom to admit fresh air. After passing over the motor, it would be emitted through another opening at the top of the box. It was not made clear how long the ventilating pipe would be, nor where the fresh air would come from, but it would not be delivered under pressure, for it was stated:

> 'there is no chance of the ventilation being reversed because the motor gets heated and with the heated air inside, you would draw the draught from the bottom to the top'[13]

This was hardly a well thought-out proposal, as fresh air would not be brought in until the motor had been running for some time and reached a high enough temperature to induce natural draught. In the meantime, sparking could have occurred and possibly ignited gas collected whilst the motor was at a standstill and cold or cooling. Snell appeared to see no problems with the idea, as he stated, 'I think we should get over the difficulty in this way entirely'.[14]

Another suggestion he thought worthy of consideration, though by no means equal to his boxed motor, had been put forward by Brogden, a mining engineer from South Wales. This was to surround the whole motor with three skins of gauze, leaving a space between each of about half an inch; if an explosion did take place, it should then be confined to the inner casing.

A.T. Snell may have thought that motors suitably protected by such means could be operated safely in-bye in the presence of gas, but his company's policy had been much more cautious. He described their three plants working underground, all specially walled-in at the intake and ventilated through a six inch pipe from the intake side. The fresh air thus admitted passed over the motors to the up-cast shaft through an open door. By this means, he said:

> 'we have not the slightest difficulty, although at times there has been sparking and I would not attempt to tell you to the contrary'.[15]

Such working arrangements bear little or no relation to those possible in-bye and were hardly a fair test of the electrical equipment's ability to operate successfully in a coal mine. There would be little or no risk of gas entering the enclosure, nor would these motors encounter the other problems commonly found in the deeper recesses of a coal mine, such as roof falls, coal dust, water, space restriction or long distribution problems, etc. It is significant that a clear disclaimer to this effect was not made.

By 1891, Snell had come to doubt the possibility of producing a motor that could be safely operated in a coal mine. At a meeting of the Institution of Civil Engineers, he told the members that he:

> 'personally was inclined to lay the utmost stress on the necessity of providing ample ventilation by well-designed fans rather than trusting to any safety device in the motor itself; for in the latter case, there would be the risk of faulty construction or accidental damage and the personal equation of the attendants must always be an important factor.'[16]

There were two ways he thought motors might be protected. The first was to enclose the whole of the armature, but that would trap so large a volume of air that, should an explosion take place within the machine, the cover would be unlikely to contain it and the gas outside would be fired. Alternatively, one might enclose the commutator only. The volume of air then involved might well be too small to do any harm but, with the windings unprotected, a short circuit might take place at any time and cause an explosion.

This rather pessimistic summary suggests that the General Electric Power & Traction Co. had abandoned the idea of a safe electric motor for fiery mines. However, his remarks might be a pointed reference to the totally enclosed machine patented and pioneered by L.B. Atkinson (joint author of the paper to which Snell was replying) in his capacity as designer for W.T. Goolden & Co., the firm marketing this type of machine in mines.

This reveals the difficult position electrical manufacturing firms were in at that time. On the one hand, they had to develop a new technology in conditions far from ideal, and on the other, they faced the sceptical or even hostile colliery management and/or the advocates of an alternative power source, not to mention the rival electrical manufacturer, waiting to capitalise on the mistake of others.

4.3 Introduction of 'flameproof' equipment

In constrast to Snell's somewhat ambivalent approach, Llewelyn B. Atkinson set about the task of providing a safe electrical installation in a far more dynamic and professional manner.

Atkinson, like Snell, was a product of King's College London, and joined the Halifax works of Messrs. Goolden & Trotter in 1884, shortly after graduating. From the outset, he demonstrated the considerable experimental and organisational skills that were to make him one of Britain's foremost electrical engineers and industrialists.

In 1886, Atkinson was introduced to mining requirements and helped to produce what he claimed to be the first electrically powered coal cutting machine.[17] The machine was made for Messrs T. & R.W. Bower and on-site testing was carried out at their Leeds colliery. Atkinson was later joined by his brother, Claude, also a King's College graduate, and comprehensive details of their work in this field over a period of some four years were given in a joint paper presented to a meeting of the Institution of Civil Engineers in February 1891.[18] The Institution awarded the Atkinson brothers the Trotter Prize for the presentation of this paper.

One way of reducing to a minimum the risks of operating electrical machines in coal mines is, as described earlier, to install motors like pumps or ventilating fans adjacent to the downcast shaft, where adequate fresh air can be provided. If motors must be sited at the coal face, like those to power coal cutting machines or rock drills, the risk from sparking commutators must be taken into account. The motor used to power the experimental coal-cutting machine tested at Leeds was of the open type, with no protection against dust or gas: it must be assumed that the colliery of T. & R.W. Bower was gas-free. After the trials, the Atkinson brothers considered the safety question and concluded that the problem resolved itself into two parts:

(i)   Would a motor working in ordinary and satisfactory conditions, at its normal and designated site where motive power is required, be likely to cause an explosion?

(ii)   Would it be possible, under extraordinary but feasible circumstances, for it to cause an explosion?

It had been noted that motors operating at voltages of up to 500 volts with their brushes correctly adjusted gave no visible sparking. If the brushes were badly adjusted or worn, slight sparking would occur. They reasoned that under these conditions the temperature rise would be insufficient to fire either coal gas or marsh gas, since the brushes were in contact with the large mass of copper forming the commutator and this would conduct the heat away. They argued that, for coal gas or marsh gas to fire, a temperature of 900°C–1,000°C would be needed, assuming mixtures of gas and air in the ratio 1 : 5 and 1 : 9 respectively at normal pressure. As copper readily oxidises at about 700°C and no oxide had been detected around the brushes or commutator, they reasoned that the ignition temperature had not been reached. They therefore concluded that, under normal conditions, an electric motor could operate safely even in atmospheres containing an explosive mixture of gas.

They thus presumed that their first condition had been satisfied but readily admitted that a fault could develop and a brush become badly worn or displaced and then temperatures might rise sufficiently to ignite inflammable mixtures of gas. The problem, it seemed to them, was how to obviate this danger. To this end, they devoted considerable attention, including many hours working underground.

Careful design, construction and operation could minimise the faults arising but not eliminate the risk altogether. Therefore, they had to assume that a machine would not normally be operated in an explosive atmosphere. Some motors, however, were left unattended for long periods, for example, dip pumps, and gas might slowly accumulate. Where motors were used at the coal face or heading there could be a sudden inrush of gas. In such circumstances, as in other emergencies, the supply needed to be cut off from the motor. This, in turn, would introduce further problems with sparks coming from the motor starter or isolator contacts, so these would need protecting.

To combat the problem, commutator and brushes and, where convenient, the whole armature were enclosed in a casing as near air-tight as possible. At that time, electrical engineers thought this was feasible only with small machines, owing to the loss of ventilation through the machine, but machines from 1 hp to 45 hp had been enclosed without encountering any difficulties. The technique employed was a simple extension of a recognised procedure. It had for some time been accepted that the temperature rise in the field winding magnets depended upon the external cooling surface area; for a temperature rise of 70°F above ambient, this area could be taken as 2·3 square inches per watt to be dissipated. Applying the same fundamental principle to the windings and magnets of the armature, the Atkinson brothers arrived at an area of about 0·7 square inches per watt for the same temperature rise. This, they claimed, was basically because the air in contact with a revolving armature is continuously moving, and so transfers heat more efficiently from the armature to the external casing. They added that the armatures electrical and magnetic components must be carefully designed.

With these precautions, Messrs. L.B. & C.W. Atkinson stated that their tests showed that, at the end of long runs, the temperature of the armature was no higher than that of the field winding magnets.

Thus far, they thought the design of the motor would be satisfactory to the electrical engineer but what of the mining engineer, afraid that gas might seep into the voids of the machine and so ignite and explode the gas outside? Their machine sought to tackle this problem. Under certain sections of the Coal Mines Regulation Act, work was not allowed to proceed where the proportion of explosive mixture exceeded about 1 : 30. It was unlikely, therefore that any electric motor would be allowed to operate for longer than two to three hours in an explosive atmosphere. On this assumption, two designs had been considered.

The first was to make the casing an air-tight box with tightly fitted joints, some packed with rubber, so that an explosive atmosphere would be unlikely to penetrate within two to three hours. If gas were to leak in, when it reached 3–4% of the volume inside the machine, the gas would ignite and be contained. The carbon dioxide produced by the reaction would tend to prevent an explosion.

If the motor were required to operate for periods in excess of two to three hours, the brothers proposed to introduce inside the casing an inexplosive mixture. The simplest approach would be to use carbon dioxide from a cylinder fixed to the side of the motor, allowing a slow but constant stream of gas to be fed into the machine. By slightly pressurising the space around the armature and commutator, this should prevent the entry of fire-damp. They observed that only 14% of carbon dioxide renders the most explosive fire-damp incombustible.

This method of protection might not always be convenient or acceptable, so they offered another motor, which followed the safety lamp in not preventing the ingress of the explosive mixture but rather precluding the egress of the flames. To achieve an adequate design, they stated:

> 'that it is necessary that the volume of the enclosed gases should be small and that the passages through which the flame can escape should be of sufficient length and the metal surfaces close enough together to cool it below the point of combustion.'[19]

This, in essence, was the design latterly known as 'flameproof' construction. It illustrates an early attempt to apply simple logic and scientific reasoning to a very difficult problem and arrived at a neat and elegant solution, even though protection was only afforded to the armature and commutator.

As stated earlier, the Atkinson brothers claimed they produced the first motor for use on a coal-cutting machine in 1887, but the previous year, L.B. Atkinson, in conjunction with W.T. Goolden and A.P. Trotter, attempted to develop a non-sparking commutator, which was essential for a machine to be used in fiery mines. Their method of reducing sparking is shown in their patent No. 8003:1886. Although the complete specification was deposited with the Patents Office only some seven months before the application for their provisional patent for the coal-cutting machine motor, no mention is made of employing their invention in machines for mine working. Powering a coal-cutting machine is one of the most arduous duties an electric motor could be set so it is difficult to imagine that patent No. 8003:1886 was taken out with no thought of its later application to the motor that must then have been under consideration. Indeed, the obituary to L.B. Atkinson in *The Engineer*,[20] states that one of his first jobs at the Halifax Works was to design, with W.H. Ravenshaw, a 6 hp motor to be built into a bar-type coal-cutting machine.

Patent No. 8003:1886 sets out the problem of the sparking commutators as being due to the self inductance of the armature coil that was being short-circuited by the brush. It sought to improve this by the use of multiple brushes. As the machine rotated, current was collected by the brush as it bridged adjacent commutator segments and as one segment moved away from the trailing edge of the brush, sparking would be prevented or reduced by a second or trailing brush keeping the retreating coil in circuit. This would be done by connecting it to the main or leading brush through an element which had a relatively high value of resistance and a low value of inductance. The patent allowed for a third brush to come into play similarly, and this was connected to the main brush through a higher resistance than the second.

A variety of brush arrangements was described, the most basic being two metal brushes in separate holders insulated from each other, and connected through a resistive component, made up from various conducting or semi-conducting materials, cut into thin strips and arranged in the form of a series grid. Other components, such as a double wound wire coil, a carbon lamp or a suitable electrolytic resistance, or secondary battery, were also suitable. Such resistive elements would be relatively large and would have to be either remote from the machine or mounted alongside or on top of the brush gear.

A development of this system was a brush made up like a sandwich, with leading and trailing plates made of brass, German silver or flexible or solid carbon of high resistance. These plates were insulated from each other, except at the end away from the commutator, to give a brush with a main path of low resistance and trailing parallel paths of higher resistance. This arrangement would be much more convenient to use. Unless specifically referred to in the descriptive text of a machine of the period, it would not be possible to tell whether new or conventional brushes were being used.

In 1887, Messrs. Atkinson & Goolden took out two patents for 'Improvements in dynamo-electric generators and motors'. In the first, No. 12,676:1887, the provisional specification is dated 19 September 1887, with the complete specification deposited on 19 June 1888. In the second, No. 16,623:1887, the dates were 2 December 1887 and 3 September 1888.

Development work must therefore have been taking place on these machines simultaneously and whilst no mention is made of the other in either patent, it is of interest to compare the two descriptions.

In patent No. 12,676:1887, the purpose is stated as being to totally enclose the armature and commutator. It was therefore intended to contain sparks coming primarily from the commutator but possibly also from the armature. The latter could arise as a result of failure of an armature coil or end connection, or from contact between the rotating body of the armature and the pole face of the field magnet, due to a bearing collapse or similar mechanical fault.

The motor was so constructed that it could be:

'safely worked and efficiently ventilated, while in the presence of explosive gases, such as obtain in coal mines, without the risks attending upon the use of such machines as heretofore constructed and when working under the like conditions.'

In contrast, patent No. 16,623:1887, makes it clear that the intention was to enclose only the brushes and commutator of a machine suitable for situations where 'it is not necessary or advisable to enclose the whole armature'.

The patent gave no hint as to what such situations could be, where it would not be necessary or advisable to enclose the whole armature, yet was presumably advisable or necessary to enclose the commutator.

It must be assumed that Atkinson & Goolden considered that there was a requirement for such a machine, and that the motor, covered by patent No. 12,676:1887, either could not fill the need or perhaps was of too high a specification and therefore too expensive. They might have thought that the output of the machine with the enclosed armature would be restricted owing to the increased temperature in the armature, caused by a lack of ventilation. This, however, seems not to have been the case as the Atkinson brothers observed (see p 69) that the size, (i.e. the mechanical output) did not appear to be a problem when enclosing both the armature and commutator: they had adopted this method with success for a motor which had developed 40 to 45 hp. Possibly by February 1891 when they presented their paper to the Institution of Civil Engineers, their development work had reached the point where they could successfully make a 40 to 45 hp machine; whereas when the patents were taken

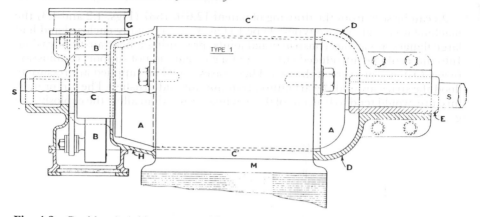

Fig. 4.2 Goolden & Atkinson patent No. 12,676:1887, Type 1
Illustration taken from patent application

out in 1887–88 they envisaged that the smaller machine would have both
armature and commutator covered, but the larger machine, only the
commutator.

In seeking to protect the commutator and armature in patent 12,676:1887, the
designers clearly recognised that both were potential sources of trouble if not
enclosed. They must also have recognised that the form of enclosure they elected
to use would also bring a number of mechanical and electrical problems.

One of the main drawbacks of this arrangement would be the recommendation
that the metal forming the protective case should be non-magnetic (gun metal
was cited as suitable) and made in the form of a tube, which the patent said:

> 'passes between the magnet bars, to each of which it is rigidly and
> air-tightly secured and thus, completely envelops the armature.'

The drawing submitted with the patent confirms this arrangement. It was
recognised at this time that the greater the gap was between the pole faces of the
field magnet and the armature, the larger the field ampere turns would have to be
to maintain the same flux density in the armature. Indeed, the precise
relationship between these quantities could be accurately predicted.[21] To
introduce into this space a cover of non-magnetic material, thick enough to
contain an explosion, would add significantly to the power required by the field
windings. They, in turn, would have to be made proportionally larger to avoid
overheating.

The mechanical problems would include obvious ones, such as the strength of
the casing to withstand explosion and the tolerances of the metal flange faces to
prevent the egress of flame.

The patent allowed for the cover to be of wire gauze, but this option was not
adopted. It will be remembered that with the traditional miner's safety lamp, the
flame was protected by wire gauze but explosions still took place. A contributing
factor in this was the velocity of ventilating air through the mine and hence
through the wire gauze. It can only be assumed that the designer considered wire
gauze less advantageous than continuous section for screens.

As can be seen from the drawing of patent 12,676:1887 (Figs. 4.2 and 4.3) the machine was made in two types. Type 1 (Fig. 4.2) was totally enclosed and had large flanges at various constructional joints, presenting a near air-tight casing. Into this could be introduced either air or a gas that did not support combustion (e.g. CO_2) or a mixture of the two. These gases would be supplied at pressure to flood the space around the armature, commutator and brushes. The air, or inert mixture, would tend to leak out of the machine rather than allow the explosive pit gas to leak in.

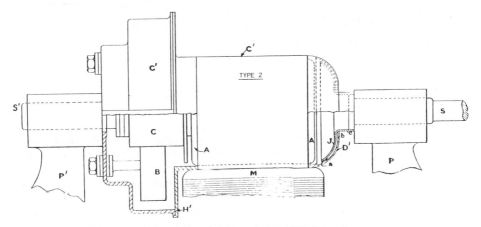

Fig. 4.3 Goolden and Atkinson patent No. 12,676:1887, Type 2
Illustration taken from patent application

In Fig. 4.3 an alternative type is shown; here the basis of design is to allow, as described in the patent:

'the free ingress of air for ventilation to such parts but will prevent the egress of flame therefrom.'

The ventilation channel, labelled *a* in the drawing, is formed by a belled end piece J, positioned relative to the end casing D, such that flames passing down the channel and emerging at aperture *b*, would be cooled to the point where they would not ignite gas outside. This channel could be further modified, as described in the patent:

'It will be understood that to further prevent the egress of flame, the adjacent faces of the dome *J* and the prolongation D^1 may be provided with a number of thin ribs, or vanes, so disposed and formed as to constitute a number of minute channels lying either in planes parallel to the axis of rotation of the shafts S, S^1 or more or less oblique thereto.'[22]

The intention here it would seem, was to increase the surface area of the metal casing to which the outward moving hot gases would be exposed in a slightly different form.

In both types, the regular adjustment of brush positions for changes in machine loading would not be simple. Indeed, with the arrangement as shown, it must be assumed that once set for a particular machine operation, the brushes were not adjusted.

Neither patent mentions providing the field winding with any form of protection, although there must have been some basic mechanical protection. There would clearly be some risk from a fault on the field winding but in machines of this period the iron work forming the poles was normally very large and cumbersome. If this awkward-shaped mass were enclosed it would leave a relatively large space in which gas could collect. Unless this space was linked with that of the armature, there would be little air movement and cooling could be a problem. If the two sections were linked, easier circulation of air may have reduced the risk of local overheating but the space liable to be filled with gas would also be enlarged, thereby increasing the strength of a potential explosion. To combat this problem, the casing would need to be very much more substantial and to have more and larger bolts holding the various sections together, as the tensile forces in the bolts resulting from an explosion might allow the flange joints to open, and increase the risk of flames escaping and igniting the gas in the mine.

A feature shared with the earlier patent (No. 12,676:1887) was the allowance for filling the chamber, at a pressure above ambient, with air or an incombustible mixture of gas. As before, CO_2 and air is mentioned. The two patents differ in their descriptions of the joints used in forming the enclosure or chamber. The basic principle, established in patent 12,676:1887, was that openings would be bounded by extended metal surfaces so that any flame or hot gases emitted would be cooled below the point where they would ignite pit gas outside. However, whilst that patent described how the machine was put together and worked, no name was given to the method of construction. The later patent No. 16,623:1887, used the words 'flame proof'.[23]

It is clear that the main problem facing the motor designer at this time was how to contain the explosion resulting from gas finding its way into a motor casing and

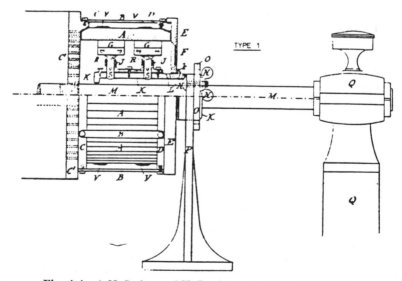

Fig. 4.4   A.H. Stokes and H. Davis's patent No. 18,206:1890

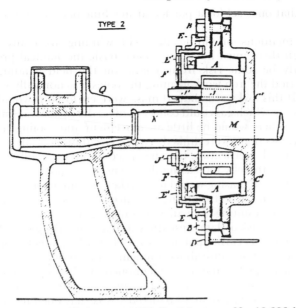

Fig. 4.5 A.H. Stokes and H. Davis's patent No. 18,206:1890
Illustration taken from patent

being ignited by sparks from the commutator. The power of the explosion would depend upon the amount of gas within the casing, so efforts were directed to reducing the internal volume of the machine.

A novel approach to this was made by Arthur H. Stokes, HM Inspector of Mines, and Henry Davis, an electrical engineer, in their patent No. 18,206;1890. In essence, their invention simply reversed the conventional mode of operation of the commutator by arranging the brushes to make contact with the inner surface of the commutator segments, rather than the outer surface. This arrangement is shown at Fig. 4.4, Type 1.

An alternative and somewhat neater arrangment using the same principle is shown at Type 2. Here, the inner clamping ring c^1 is screwed or keyed on to the shaft M, to form a permanent closed end to the commutator cylinder; at the same time it abuts onto the armature.

The T-shaped commutator segments A are clamped into position by the ring D and held in place by bolts B; the heavy line around A represents an insulating material (vulcanised fibre). The fixed outer face of the disc E′ is held in place by the locking ring E, which in turn holds in place the retaining bolts B by an external upper edge; the clamping bolts were thus prevented from vibrating loose. The locking ring E might work loose, so machines rotating in a clockwise direction (viewed from the commutator end), had a ring with a left-hand thread and machines with a counter-clockwise motion, a right-hand thread. The disc E′ fits closely but sliding around the shaft sleeve K, giving flame-tight joints at the inner and outer edges. If not restrained, there would be a tendency for the ring E to rotate when the machine was in motion, owing to the closeness of the fit at the outer edge. A restraining force is provided by the locking handle Y, passing

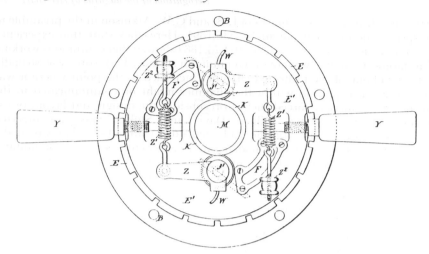

Fig 4.6   A.H. Stokes and H. Davis's patent No. 18,206:1890. Type 2 brush gear assembly
Illustration taken from patent

through lugs on disc E' and tightening on to the sleeve K. Adjustment of the
brushes J was achieved by slackening off the locking handle Y and rotating the
disc E', complete with brush assembly.

Fig. 4.6 gives a clearer view of the handles and brush assembly, showing how
pressure on the brushes was increased or diminished by adjustment of the milled
nut Z^2.

Referring to this Davis and Stokes invention, A.T. Snell observed[24] that, whilst
it had advantages, the mechanical difficulties involved in handling it were
considerably increased. His company was at that time building a 25 hp motor
with an inverted commutator and were having some difficulties with the internal
brushes. He hoped to get over this problem by using carbon brushes.

In spite of a number of drawbacks, the Davis and Stokes inverted commutator
enjoyed a limited commercial success as two contemporary sources show. A.T.
Snell observed in the second edition of his book[25] that:

> 'Experience extended over more than eight years has demonstrated
> that gas fired inside the commutator is not communicated to the
> outside.'

C.W. de Grave told fellow members of the Association of Mining Electrical
Engineers in 1920[26] that, although a number of these machines had been sold
and on these the systems worked well, the idea never really caught on. The main
reasons, he thought, were that the commutator overheated and the system only
eliminated risks from sparking brushes. The arrangement had been thoroughly
tested: even at high velocities of explosive mixtures and with gas burning inside
the commutator, gas outside the unit could not be fired. Disregarding its limited
commercial success, he thought it had served a useful purpose in demonstrating
that flame-tight chambers could be produced.

The problems facing the mining motor designer during this early period are

further illustrated by the remarks of L.B. and C.W. Atkinson in the preamble to
the specification in their patent No.536:1891. Here, they state that experience
with motors designed to patent No. 16,623:1887 showed that continuous working
of the motor resulted in wear of the commutator surface. This wore away so badly
that the volume of the protected chamber increased to the point where it was
feared that an explosion within the chamber might be communicated to the
outside. Their anxiety forced them to reconsider the design and layout of the
conventional commutator. Where Davis and Stokes chose an inverted
commutator, the Atkinson brothers adopted a disc commutator that was rigidly
attached to the drive shaft at right angles to the axis of rotation. The commutator
segments were mounted on the outer face of disc; the method of fixing and
insulating is not stated in the patent. A dish-shaped, close-fitting cover carried
the brushes in specially formed recesses; this cover was held in place by a locking
ring that allowed the commutator disc to rotate while maintaining a close-fitting
flame-tight seal. The claim was made that with this design as the commutator
strip wore, the outer casing could be adjusted, taking it nearer to the commutator
and thus keeping the internal volume to a minimum.

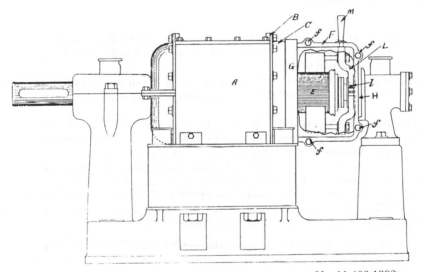

Fig. 4.7   W.T. Goolden and L.B. Atkinson's patent No. 11,403:1892
Illustration taken from patent

Unlike the Davis and Stokes inverted commutator, there appears to be no
evidence to show that the Atkinson brothers' patent was taken up commercially.
When Atkinson and Goolden introduced their latest machine in 1892 (patent No.
11,403), it was fitted with a conventional commutator.

The machines of W.T. Goolden & Co. covered by patents Nos. 12,676:1887
and 16,623:1887 must be considered as prototypes used by the designers to
explore previously uncharted ground. Both patents had features that imposed
limitations as well as others that were a clear advantage. In 1892, Atkinson and

Goolden took out patent No. 11,403 (Figs. 4.7 and 4.8 *a–c*) which was an improvement on the two earlier patents, extending a number of the principles previously introduced.

The most notable improvement was the use of the pole faces as part of the commutator enclosure, obviating the necessity for a non-magnetic tube, previously interposed between the armature and the pole pieces. It will be seen in the part section at Fig. 4.8*b* that, at the top and bottom of the machine, the gaps between the two pole pieces are closed by metal plates with side flanges. The side elevation at Fig. 4.7 shows the casing end plates, also bolted to the pole pieces. At the drive shaft end, the cover is formed by two castings, shaped to fit closely over

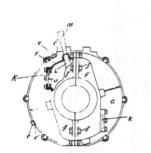

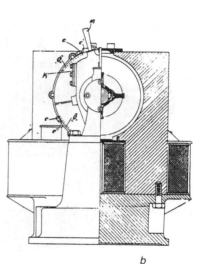

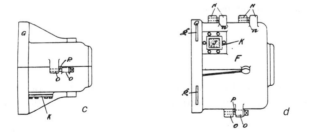

Fig. 4.8   W.T. Goolden and L.B. Atkinson's patent No. 11,403:1892
Illustrations taken from patent

the end of the armature. These plates are bolted together through wide flanges. It must be assumed that the two plates are recessed into the metal work of the pedestal bearing, again making a flameproof seal, but the patent is not explicit on this. It does not say whether ventilation ducts of the type employed in patent No. 12,676:1887 were used or not.

At the commutator end of the machine, the commutator cover F is sealed at its bearing end by a wide pivotal journal on bush H. The end plate B, is bolted to the pole piece and so shaped that it presents a cylindrical profile to the cover F; the two covers are meant to have a close overlapping fit. Thus the whole commutator cover is free to rotate, allowing for adjustment of the brushes, and, once set, the two covers would be bolted together by the four bolts *c*, through adjustment slots *g*. It will be seen from the end elevation at Fig. 4.8*a* that the cover F is near rectangular in shape, where it covers the commutator, so the internal volume is kept to a minimum whilst allowing for the mounting of the brushes.

Fig. 4.9 Goolden safety mining motor
Illustration taken from *Electric motive power* by A.T. Snell, p 339 (shown in both 1894 and 1899 editions)

The patent says that, as this type of machine would be expected to work continuously for long periods, it would be necessary to inspect the brushes and commutator three or four times a day. This had been allowed for in the design as the commutator cover could simply be opened by the removal of the fixing bolts *f* and *c*. Alternatively, a hinged cover was provided (Fig. 4.8*c*). Here, the two sections are hinged at N and drawn together by bolt P. The fixing bolts *c* are still retained to ensure that the cover could not rotate and misalign the brushes. It would also, of course, ensure that the gap between faces C and G was kept to a minimum. It will be noted that the field windings are still not included within the 'flameproof' cover.

As pointed out earlier, this motor was fitted with a conventional commutator rather than the type covered by patent No. 536:1891 or some other design with reduced volume in the commutator chamber. From this it would appear that

Fig. 4.10   Endless rope haulage plant driven by Goolden safety motor
Illustration taken from *Electric Motive Power* by A.T. Snell p. 328 (shown in both 1894 and 1899 editions)

during the period when the design for the modified commutators was being considered, (application for patent 536:1891 was made in January 1891 and accepted the following month) to the date of the submission of the patent for the new motor (patent No. 11,403:1892 was applied for in June and accepted July 1892), the thinking on motor design went through a fundamental change.

From contemporary literature, it would seem that this Atkinson and Goolden machine remained on the market for a number of years. Snell, in his book of 1894, says[27] it is 'an excellent safety mining motor', and illustrates the motor complete with double helical gear drive wheel (Fig. 4.10). A further example of this type of motor is shown in Snell's book (Fig. 4.11). Here the motor is fitted with external fins, to increase the surface area and aid cooling. Figs. 4.10 and 11 show that the interconnecting cables on the motors were given no protection from mechanical damage. If these cables were broken or pulled out from the terminals, arcing might result, with obvious consequences if gas was present at dangerous concentrations.

In the second edition of Snell's book, published in 1899, the motor is again illustrated, but this time his comments are less complimentary. Referring to the Atkinson and Goolden machine and a number of others then being manufactured, he says:

> 'It is questionable, however, whether all these designs are not chiefly important as dust protectors. (It is absolutely necessary that a motor worked on a coal face should be dust proof). Though the cases may fit well when new, they probably soon get loose and defective from rough usage inseparable from the class of work and thus become useless as safety devices; indeed, they may then be a positive source of danger from the apparent security attributed to them.'[28]

Attention had therefore recently been turned to the so-called 'enclosed' motor, where the casing enclosed the whole machine, protecting the field windings as

Fig. 4.11 25 bhp Thames Iron Works & Shipbuilding Co. enclosed motor
Illustration taken from *Electric Motive Power* by A.T. Snell, 1899 edition p. 341

well as the armature and commutator. He gave, as an illustration, details of a machine of which he was a co-designer. (Fig. 4.11).[29] As can be seen, the enclosure for this machine was simply a rectangular tank, made from cast steel and of large section with wide flanges. The free air space within this enclosure must have been considerable and, to contain any possible explosion, the casing would need to be substantial.

Also illustrated in Snell's second edition is a drawing of Mavor and Coulson's steel-clad mining motor.[30] This is an example of the new generation of machines then emerging, which took advantage of the recent developments both in the design of 'flameproof' enclosures and of machine design. In spite of these and other improvements Snell concluded the section of his book on *Motors and Driven Machines* with the comment that, in his view:

> 'the only real safeguard against explosion from sparking brushes and commutators is to do away with them altogether and the only way in which this is practicable at present is by the use of the polyphase motor'.[31]

In the second edition of his book published in 1899, the above view is restated.[32] Thus, as early as 1894, Snell saw the motor of the future for mine working as an ac machine, in spite of the advantage of dc in the ease of controlling speed and torque.

4.4 The wider view of testing and development

By 1914, a great deal of investigative work had been carried out on the ignition of firedamp and coal dust by electrical apparatus. Although more was understood about the phenomenon and the design of motors and switchgear had been greatly improved, under certain circumstances the equipment still offered little or no protection.

The early experiments of engineers such as Atkinson and Snell and the reports of the Prussian and French Commissions, indicated that under certain circumstances electrical apparatus could explode gas. However, in 1898, Bergassessor Heise and Dr Thiem published their more authoritative report on tests carried out at the Gelsenkirchen firedamp station.[33] This report covered extensive experiments into the ignition of firedamp and coal dust by a number of electrical devices, including incandescent filament lamps, arc lamps, switches, safety fuses, resistances and ac and dc motors.

The results on motors showed that sparking of the commutators of small dc motors could easily ignite firedamp. However, with large motors in perfect working order, explosions were not produced under test conditions. To some extent, this result backed up the earlier views of Snell and the Atkinson brothers. The report concluded that, as ideal conditions are rarely found in underground workings, it would be preferable to:

> 'enclose either the entire motor or those parts of it that produce sparks in a flameproof casing.'[34]

Their experiments indicated that ac motors were 'devoid of danger'.[35]

From their tests on switches, they noted that the strength of current broken by the switch gave no measure of the danger of the spark produced. Difficulties arose when voltage and self-inductance of the circuit connected to the switch were considered, and they concluded that:

> 'it seems impossible to define the limits within which the sparks produced on breaking contact are, or are not, dangerous.'[36]

It can only be assumed that this finding left the electrical equipment designer wondering how the problem could best be tackled, and at the same time provided useful ammunition for those opposed to the introduction of electrical plant in mining.

During the first series of experiments at Gelsenkirchen, no tests were carried out on any form of protected motor, but in 1903 the tests were extended to cover this type of equipment. It was later reported[37] that motors protected by wire gauze were unsatisfactory due to afterburning[38] and that the best form of protection was obtained by providing ventilation through a series of laminated plates. These plates, usually two or three in number and made of brass or other non-rusting material, were normally built into the top of the motor housing (the stator) or made to form part of the end plates.

Following an exhaustive series of tests, the US Bureau of Mines came to similar conclusions and published its results as a technical bulletin in 1912.[39] The Bureau tested five types of safety device attached to motors; A and B, had gauze covered openings, C and D used baffle plates, and E a combination of gauze and baffle plates.

Of the two gauze-protected motors, type B was rejected without testing and type A, declared unsatisfactory after repeatedly discharging flame through the protective gauze in 187 times out of 191 tests. Of the two baffle plate protected devices tested, the type C consisted of five brass plates spaced ¼ inch apart and had a free air path of 28 inches. The obvious fault was that the baffle plates were spaced too far apart.

The second plain baffle plate protected motor, type D, was fitted with an internal fan. Separate tests were carried out on the motor with and without the fan in operation. Without the fan, all the tests were completed satisfactorily and there were no signs of flame being discharged through the plates. With the fan in operation, flames were discharged through the plates and afterburning occurred. This device was also rejected.

The final device, Type E, was subjected to 272 internal gas explosions and at no time was flame seen to be discharged from the plates. They concluded that the device gave adequate protection in the presence of gas alone. When dust was present in the device, however, it tended to ignite. Under these conditions, the device did not afford the protection it was designed to give.

These tests were carried out by the Bureau as part of a series to establish a 'permissible list' of machines for use in mining, but the bulletin did not say whether these five represented current American practice. This may be inferred, however, because a number of firms in the USA were at that time willing to submit their equipment to independent test, presumably in the hope of being put on the approved list, of the five types tested, non were given fully satisfactory clearance.

There was no similar government department testing or approving equipment in Britain but some comparison can be drawn between practice in Britain and the USA from an article by T. Duckitt of the General Electric Co. This appeared in *The Electrician*, Second Mining Issue, in 1911[40] (the first had been in 1908).

Duckitt says that most motors for mining applications fit one of two categories:

(a)   Totally enclosed
(b)   Gauze enclosed

There were also many designs where only parts of the motor, such as the commutator, were protected.

His comments regarding the gauze enclosed motor generally agree with the findings of the US Bureau of Mines. Owing to this type's unreliability, laminated plate covers or guards, which he said originated in Germany, were then coming onto the market.

The organised testing of mining electrical equipment in Britain at this time appears to have been some way behind that of Germany and the USA. The work in Britain was carried out in a largely unco-ordinated manner by manufacturers, colliery engineers or academics.

W.E. Garforth reported a series of experiments to the first Departmental Committee on the Use of Electricity in Mines, in 1903.[41] They indicated clearly that gauze protected dc motors could not be relied upon in gassy atmospheres.

Details of tests carried out by a motor manufacturer, J.H. Holmes of Newcastle upon Tyne, were reported to the second Departmental Committee in 1910.[42] These tests, carried out with some assistance from Professor Bedson of the Armstrong College, were intended to check if a ventilated flameproof motor could be made that would be safe under all conditions.

The main series of tests used motors ventilated through bundles of plates, following the earlier German tests. They found they could not guarantee the safety of the device under all conditions. Their results follow closely those of the US Bureau of Mines, except that their tests went further with regard to plate size and spacing. In this context, their findings appeared to agree with those of the German tests: the most effective plate gap was of the order of 0.5 millimetres.

The committee were also told that tests had recently started at J.H. Holmes' on flameproof switch boxes with wide roughly machined flanges. These tests were not complete but it appeared that flange widths of less than one inch were unsatisfactory.[43]

An example of the totally enclosed motor was the explosion-proof type. This machine was enclosed in a metal case, strong enough to withstand an internal explosion, and was then being put forward by many manufacturers, even though the lack of ventilation could derate the output by up to 50%. In situations where gas was prevalent, Talbot Duckitt thought that this type of motor was the only reliable form to adopt.

With regard to mining switchgear, one of the pioneering firms in Britain was A. Reyrolle & Co. of Hebburn on Tyne. By 1911, this company was also making 'explosion-proof' equipment[44] and their enclosures were designed to withstand an internal pressure of 200 lb per square inch.[45]

Another leading Tyneside company, Messrs. Ernest Scott & Mountain, was also at that time making explosion-proof equipment and had manufactured a motor for a coal cutting machine which had covers with planed and scraped flange faces.[46]

Even with the new 'explosion-proof' equipment, a number of electrical engineers had serious doubts about the use of electrical equipment in mines where high concentrations of gas were likely.[47]

This matter was raised in the House of Commons in 1911, during the passage of the Coal Mines Bill, when Atherly-Jones KC stated that he

> 'was perfectly convinced that the coal-getting machines worked by electricity were sources of the gravest danger.'[48]

He also said that he hoped the Bill would give specific assurances on how to deal with the matter.

These general fears were reflected in the findings of the Departmental Committee appointed by the Home Secretary in October 1909.

The report was critical of many aspects of design and construction of colliery installations and equipment and the Committee felt that cheapness, rather than suitability for the task, was too often the main criterion. This tendency they thought, was made worse by the intense competition among manufacturers of electrical apparatus. However, in spite of these misgivings on the quality of equipment, the committee reported that it could not recommend that the Home Office should give official type or design approval to equipment.

It was clearly not thought desirable that Government should get involved too closely with manufacturers' designs and specifications for equipment as in the USA. Instead, they sought to improve the standard of installation and design by strengthening the Special Rules (see chapter 7).

4.5 Notes and references

1   *Transactions North of England Institute of Mining and Mechanical Engineers*, 1891–92, Vol. 41, p. 178
2   *Proceedings Midland Institute of Mining, Civil & Mechanical Engineers*, 1887–89, Vol. 11, p. 344
3   *ibid.*, p. 342
4   *ibid.*, p. 368

5 *ibid.*, p. 317
6 *ibid.*, p. 353
7 *Proceedings Institution of Civil Engineers* 1887–88, Vol. 91. p. 108
8 *Trans NEIMME*, 1852–53, Vol. 1, p. 282
9 *Proc MIMCME*, 1887–89, Vol. 11, p. 354
10 About this time, this problem was being tackled by ensuring that the motors ran lightly
   loaded, i.e. the machine was derated. Some improvements were also made by the use of
   carbon brushes and by modifications to the general design.
11 *ibid.*, p. 347
12 *ibid.*,
13 *ibid.*, p. 348
14 *ibid.*
15 *ibid.*
16 *Proc ICE*, 1880–91, Vol. 104, Part 2, p. 118
17 *ibid.*, p. 99
18 'Electric mining machinery, with special reference to the application of electricity to
   coal cutting, pumping and rock drilling,' *ibid.*, p. 89
19 *ibid.*, p. 95
20 *The Engineer*, 1939, Vol. 168, p. 186
21 See Thompson, S.P.: '*Dynamo-Electric Machinery*', 3rd Edition, 1888, pp. 311–325,
   where reference is made to J. and E. Hopkinson's paper to the Royal Society in 1886
   and to Professor Forbes's three propositions or lemmas. On p. 323, S.P. Thompson
   states that, according to Ravenshaw, Mr Forbes's lemmas were used recently by
   Messrs. W.T. Goolden & Co. and 'they enable the designer to predetermine the
   performance of the machine within two percent.'
22 From the foregoing, it is clear that for these motors to be successful, some accurate
   machining would be needed. Rolt indicates that close tolerances were being achieved
   in the 1880s where 'the engineer used a thousandth of an inch as his yardstick.' See
   Rolt, L.T.C.: *Tools for the job;* Batsford, London, 1965, p. 187
23 This is the earliest use of the word 'flameproof' that has been found in the reference
   material consulted. The term was eventually accepted by the Home Office and later by
   the British Standards Institution. In the meantime, other terms such as
   'explosion-proof', 'flame-tight' or 'enclosed' were frequently used, though exact
   flameproof definitions for these terms had not been agreed.
24 *Proc ICE*, 1890–91, Vol. 104, Part II, p. 118
25 SNELL, A.T.: '*Electric motive power*', 1899, p. 340
26 *Proceeding Association of Mining Electrical Engineers*, 1910, Vol. 1, p. 235
27 SNELL, A.T. 1894, p. 340
28 *ibid.*, 2nd Edition 1899, p. 342
29 *ibid.*, p. 341
30 *ibid.*, p. 343
31 *ibid.*, 1st Edition 1894, p. 342
32 *ibid.*, 2nd Edition 1899, p. 342
33 'Experiments of the ignition of fire damp and coal dust by means of electricity', *Trans
   IME*, 1898–89, Vol. 17, p. 88
34 *ibid.*, p. 116
35 *ibid.*
36 *ibid.*, p. 105
37 *Proc AMEE*, 1910, Vol. 1, p. 232
38 Afterburning is the term for the condition where the gas continues to burn after firing
   so that the gauze is damaged or destroyed.
39 'An investigation of explosion-proof motors, *Technical Bulletin* No. 46, US Bureau of
   Mines, 1912
40 DUCKITT, T. 'The gas-tight and flameproof motor problem'. *The Electrician* Second
   Mining issue, 12 May 1911, p. 35
41 *Report of the Departmental Committee on the use of Electricity in Mines*, HMSO 1904, Cd. 1916,
   Q.6529–44
42 *ibid.*, 1911 Cd 5498 Q4604–95
43 *ibid.*, Q4675

44 The term 'explosion-proof' came into general use about this time to differentiate between the two classes of 'flameproof' equipment (*a*) that was 'gas-tight' and had covers fitted with gaskets which tended to be blown out by an explosion and (*b*) that had covers with wide machined flange faces

45 CLOTHIER, W.H.: *Switchgear stages*, 1933, p. 57

46 *Proc AMEE*, 1910, Vol. 1, p. 233

47 *a Proc AMEE* 1910, Vol. 1, p. 234
   b The Electrician Second Mining Issue, 1911, p. 42

48 *The Electrician*, 1911, Vol. 66, p. 963

Chapter 5
Coal face mechanisation

5.1 Mechanisation and the demand for greater output

Over the period 1850 to 1913, coal output in the UK increased fourfold (see Fig. 5.1). The Cambridge economist, B.R. Mitchell, argues that the increased demand for coal in the second half of the nineteenth century was met in Britain in two main ways: firstly, by developing new reserves, i.e. opening up new pits and new coalfields, and secondly, by exploiting reserves better with improved mining technology.[1]

The improvements in mining technology introduced at this time cover a wide

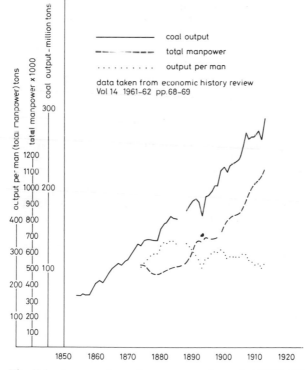

Fig. 5.1   Increase in coal output over the period 1850–1913

field, from improved methods of winning coal to new mechanised processes replacing traditional methods of operation.

One of the most important developments was the gradual change from pillar and stall working to the longwall system. The advantages of long wall working according to Caleb Pamely[2] were essentially that, where it was possible, it was more efficient and cost-effective, particularly in deep mines where the weight of rock above tends, with favourable roof conditions, to reduce the amount of undercutting needed to bring down the face. This system provides better ventilation and results in more of the coal being removed; it lends itself better to machine working and offers easier face haulage routes.

Over this period, mechanised systems were improved for: steam winding for coal, men and materials, the manufacture and use of wire ropes for haulage and winding[3], ventilation and fan design, main road haulage, and water pumping.[4]

A notable exception was the application of mechanised techniques to coal cutting. The overwhelming majority of men working at the coal face still relied on traditional methods of hewing coal, assisted in the later part of the century in certain areas by blasting to bring down the face.[5] Highlighting this lack of development of face mechanisation and the introduction of electricity underground, N.K. Buxton claims that innovation was markedly slower in Britain than in the USA or Germany. The full effects of 'this relative decline in technical standards' was not fully felt until after the 1914–18 war, when the British coal industry was confronted with new and unfavourable market conditions.[6] As can be seen from Fig. 5.1, from 1874 to 1913 the output of coal more than doubled, as did the total number of men employed, but the output per man underground never regained the peak of 333 tons reached in 1883. Mine owners' preference for traditional coal hewing techniques did not deter a small number of enterprising engineers, together with a few entrepreneurs and more adventurous colliery owners and managers, from experimenting with and slowly developing a wide range of coal-cutting equipment.

5.2 Early forms of coal cutting machines

The mechanisation of coal production is ultimately centred around coal cutting and there are many recorded attempts to find a substitute for human muscle power and the pick.

The earliest proposal to mechanise the cutting of coal appears to have been in 1761, when Michael Menzies, of Newcastle upon Tyne, took out a patent for a machine powered by a steam engine, or 'fire engine'. The patent also provides for power to be supplied through a 'water miln', a 'wind miln' or a 'horse gin', if more convenient. The intention was to provide motive power at the surface of the mine and transmit it down the shaft to the coal face, through an arrangement of reciprocating spears and chains, to move heavy iron picks that would hack out the coal.[7]

Menzies' patent was described by William Firth of Leeds in a paper for the British Association meeting at Bradford in 1873. Entitled "Firths Coal-Cutting Machine', the paper described a compressed air powered unit, commonly referred to as the 'Iron Man', which, according to S.F. Walker, was used at West Ardsley Colliery, Leeds, as early as 1862 and at the Hetton Colliery, Co. Durham in 1864.[8]

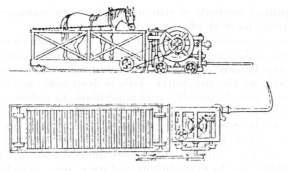

Fig. 5.2   Horse driven coal cutting machine assumed to have been used about 1860 (Source: *M & C Machine Mining*, Vol. 2, 1922)

In 1867, Theo Wood Bunning, the Secretary of the North of England Institute of Mining and Mechanical Engineers, catalogued the progress of coal cutting machines up to that time and outlined some 106 patents, ranging from simple man-powered battering rams through to steam, hydraulic, compressed air and electrically powered units.[9]

A novel approach to the problem, not mentioned by Bunning, is the horse-driven machine shown at Fig. 5.2. This is said to have been used at the Waterloo Colliery near Leeds and was a variant of the 'Iron Man'.[10]

The more successful of the early patents were powered by compressed air and fell into four categories:

- The swinging pick type is the earliest coal cutting machine, simply copying the reciprocating action of the miner hewing coal. This was the basis of Menzies' design and was copied by Firth with the Iron Man but with two pick heads on the swinging arm. The early electrically powered machines also chose a reciprocating action.
- The chain cutter: an early patent was taken out by William Peace of Wigan in 1853, though the first practical machine was built by William Baird in Glasgow in 1864. This was known as the Gartsherrie, and an improved version was introduced to the mining engineers of the USA at the 1876 Philadelphia Exhibition.
- The bar cutter appears to have been first patented in 1856 by Johnson and Dixon but it is doubtful if this machine was ever made. The successful development of this type of machine followed its reintroduction by Bower and Blackburn in 1885, described in more detail on p. 96.
- A number of rather crude disc cutting machines were made during the 1850s and 1860s, including a hand powered machine shown at the Paris Exhibition of 1851. The first practical disc cutter, however, appears to have been built by J. Scarisbrick Walker of Wigan in 1869 and used at the Ladies Lane Colliery, there. This type of machine was also used for many years in the Cheshire salt mines.

A fifth category, the percussive drilling or cutting machine, was mainly developed in the USA and introduced into Britain late in the nineteenth century. Although a large number of patents were taken out in Britain between about 1880

Fig. 5.3   Hand holing in the Silkstone seam, *c.* 1900
(Source: Royal Commission on Coal Supplies, 1st Report 1903)

and 1900 under the heading of Rock Drills (a number of which were electrically powered), few were used in British coalfields during this early period.

All of these early machines, and their successors well into the twentieth century, copied the miners' method of working, whether by longwall or pillar and stall systems. They undercut the coal face to a depth of some three to five feet, depending on machine type and face condition, the coal was then broken into pieces that could be easily handled. At that time, there was little or no demand for 'small coal', so the section of coal removed in undercutting by the miner or the machine had little value and it was often left underground as packing in the goaf or, at best, used in the colliery's own boiler plant. The miner would try to keep this waste to a minimum by undercutting or holing in dirt bands often present in the coal strata, whether at floor or other level. If, however, the solid coal was sandwiched between rock strata, then with any method of working, coal would have to be removed and some waste incurred.

Fig. 5.3 shows a miner having to remove a substantial section of good coal before the face can be dropped. The waste will be considerable even though the seam is a reasonable thickness at some four feet. The narrower the seam, the less economic it would be to remove a substantial wedge of coal that could not be sold.

The alternative put forward by the advocates of machine undercutting is shown in Fig. 5.4, where an early compressed air powered, disc cutting machine is shown at the start of a cut, with the cutting disc about one foot into its normal cutting depth. W.E. Garforth claimed[11] that in the early 1890s, by using a disc cutter that removed a 5 inch slot to a depth of some 4½ feet, he was able to increase the output per man per eight-hour shift from 3¼ tons by hand to 4½ tons. Later, using an improved machine cutting to a depth of 5½ feet, the average output per man was increased to 6 tons.

The question of waste coal was such an important issue that a separate sub report was produced by the 1871 Royal Commission on Coal, part of which states:

'At present, under favourable conditions of working, the ordinary and unavoidable loss is about 10 percent, whilst in a large number of

Fig. 5.4 Compressed air powered disc coal cutting machine at the start of a long wall face, *c.* 1900
(Source: *Transactions IME*, Vol. 23. 1901–02)

instances when the system of working practised is not suited to the peculiarities of the seam, the ordinary waste and loss amounts to sometimes as much as 40 percent.'

The principal cause of waste proved to be pillars left and crushed or seams abandoned or left to be crushed for whatever reason. Other causes of serious waste included:

'Much small coal is still either left below ground, or consumed in large heaps on the surface [and] the requisite holing or undermining is frequently wastefully made in good coal'[12]

In 1872 Robert Winstanley read a paper to the Institution of Mechanical Engineers entitled 'Description of a Coal Cutting machine with Rotary Cutter, worked by Compressed Air'.[13] He described a disc-type machine that he and a man named Baker had designed. In recommending the use of coal cutting machinery, he observed that:

'At no time in the history of the coal trade has a greater want been felt than at the present time for the substitution of machinery in place of manual labour in the working of coal.'

Winstanley reported that the machine had been used for nearly two years at the Platt Lane Colliery, Wigan, in the 'Pemberton Little Coal' seam. This seam was about 2 feet 4 inches thick and the coal was so hard that the management had the utmost difficulty in obtaining men to work it. For a time it had stood idle. When willing hands were found, they had to be paid a higher rate than that for working other seams in the pit.

In the discussion that followed Winstanley's paper, E. Fidler, from Platt Lane Colliery, confirmed the success of the machine. He said that in this particular seam a man would undercut by hand a section of face some 4½ yards long to a depth of 2 ft 8 in in 10 hours. During this operation, he would take out a wedge section one foot high at the face, tapering to a point at the back, and this coal would be slack. By comparison, the machine, with an air pressure as low as 12 to 15 pounds per square inch, could do the same length of cut to a depth of 2 ft 10 in in 8 minutes, making only a 3 in high cut.

In 1890 George Blake Walker gave a paper to the Federated Institute of Mining Engineers on 'Coal getting by machinery'.[14] He outlined the current 'state of the art' in coal cutting machinery, detailing performance and costs of a range of machines for compressed air or electrical drives. At the same meeting, Albion T. Snell gave a paper on 'Electricity as applied to Mining Operations', detailing the characteristics of series, shunt and compound wound machines. He also outlined different methods of distribution suitable for supplying fans, pumps, coal cutting machines and rock drills.[15]

This talk was clearly pitched at men with a fair understanding of electrical engineering. In contrast, he had given a similar talk to a meeting of the Lancashire branch of the National Association of Colliery Managers at Wigan some three months earlier but in a more elementary manner.

Thus, in the last half of the nineteenth century, the need for coal cutting machines in the British coal industry was recognised and a number of engineers were keen to take up the challenge. These engineers were not slow to inform the industry and wider interest groups of the potential of machines. Following the discovery of the principal of self-excitation and the development and expanded use of the dc machine, a small group of engineers saw the advantages of harnessing this new power source to drive coal cutting machines. However, the majority of the coal owners were less enthusiastic and the introduction of new coal cutting techniques into the British coal industry proved painfully slow.

5.3   Introduction of electrically powered coal cutting machines

The earliest British patent seeking to apply electric power to a coal cutting machine is No. 2,327:1863, taken out by Robert Ridley of Leeds and James Grafton Jones of Pentonville. The layout of this machine is shown at Fig. 5.5.

As can be seen, the machine uses a single swinging pick, whose reciprocating motion is derived from four electromagnetic coils. The semi-circular bar A passes through the centre of coils G and H and is fixed to the central axis C by the cranked arms A'A'. Similarly, the semi-circular bar B, located at a convenient distance below the A, A', A' assembly passes through coils I and J and is fixed to the shaft C by arms B'B'. Both semi-circular bars A and B are made up from alternative sections of soft iron and 'non-conducting materials'.

As shown in the Figure, the machine is ready to start a working stroke and movement is produced by first passing an electric current through coil G. The patent states:

> 'the bar A will be moved in the direction shown by the arrow and when the bar A has been moved a distance so as to bring the next piece of soft iron to coil B (*sic*)[16] the electric current is to be cut off from the

coil and contacts made for passing an electric current through the wire of coil H.'

Having completed the working stroke, power is then applied to coils I and J in turn and the pick is returned to its original position, ready to start the sequence again.

While the patent specifies that the bars A and B are made up from alternate sections of soft iron and non-magnetic material, it does not state how many sections of each. From the drawing supplied with the patent, there appear to be four sections of unequal length in each bar. If this were the case, it would be difficult to see how the machine worked. Presumably there were six sections in each arm of equal length, similar to that of the pitch of the coils. The sections numbered 2 and 4 in Fig. 5.5 would be made of soft iron and, as a current is passed through coil G, section 3 is not affected by the magnetic field but section 2 will be drawn into the coil. At the point where section 2 is completely covered by coil G, section 4 is just about to enter coil H and, at this moment, power is transferred to this coil and the power stroke completed. The bar B would then operate in a similar manner and return the pick to its starting point.

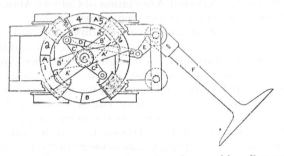

Fig. 5.5 R. Ridley and J.G. Jones electric coal cutting machine. Patent No. 2,327:1863 Illustration taken from *M. & C. Machine Mining*, Vol. 3, p. 53

The patent does not make clear how current was to be switched from one coil to the other — 'the same being well understood' — but it states that the coils were to be 'in communication with an electric or magnetic battery or batteries'. At this period, an electric battery would almost certainly have meant a primary cell. Although Planté had produced his secondary cell in 1859, it found little industrial application until 1881, following Faure's introduction of the preformed grid.[17] The magnetic battery would imply the use of a magneto-electric generator (the dynamo had still to be introduced). From either source, the power available would be somewhat limited and there would be problems with providing adequate and reliable cables from the generator, sited on the surface, down to the coal face.

It is not clear whether or not this machine was constructed and trials carried out; probably it simply joined the many other early patents for coal-cutting equipment lying in the records of the Patents Office.

Some early electrical pioneers had broad interests, like Henry Wilde who turned his attention to coal cutting equipment, following his work on the development of the dynamo and self-excitation in 1866–67. In May 1874, Wilde applied for patent No. 1554, 'Improvements in Machinery and Apparatus for Excavating Coal and other Minerals and in the Mode of Actuating such Machinery.' This provisional patent carries a note stating:

> 'Void of reason of the patentee having neglected to file a specification in pursuance of the conditions of the Letters Patent.'

In the preamble, Wilde stated that electricity was to substitute for compressed air or other systems then employed and that, as a power source, he would prefer his multi-armature machine described in patent No. 842:1867. The coal cutting machine he described worked with a reciprocating action, although he said it could also be applied to machines with rotating cutters or reciprocating motion converted from rotating armatures.

The action of the machine copied that of the miner's pick. The single bladed tool was driven on the power stroke by the force of an electro-magnet and the return stroke was accomplished with the aid of a spring. Continuity of reciprocating action was to be maintained by two sets of electrical contacts. On the completion of the power stroke, a circuit was made that short-circuited the current passing through the electro-magnets. As they were de-energised, the spring returned the arm to the starting position, where a further pair of contacts was made, allowing current to flow through the magnets again, and the cycle was repeated. Brass studs projected slightly above the surface of the iron cores of the electro-magnets to prevent the armatures coming into contact with these magnets; this helped their de-magnetisation at the end of the power stroke. A coil of a few turns was also brought into circuit, which wound round the main coil but in the opposite direction. After each stroke, the pick would be automatically advanced into the seam by means of a pawl and ratchet wheel. Provision was also made for raising and lowering the pick and for moving it along the coal face.

It is not clear why Henry Wilde did not pursue his patent further. Presumably he found the problems of constructing the machine more difficult than expected or the difficulties of providing the necessary power to the machine insuperable. For whatever reason, he abandoned the idea and in September 1874 took out a further patent (No. 3000), for a man-powered, reciprocating coal-cutting machine, again using a single pick. This time he filed a specification, with a set of detailed drawings.

For the next few years, while arc lighting was steadily introduced in industrial premises and public thoroughfares, no further attempts were made, in Britain at least, to apply electricity to coal cutting machines.

R. Bolton states that in 1880 Charles Ball of New York invented an electrically powered, percussive rock drill. This machine had two solenoids through which current passed alternately, causing a soft iron bar to vibrate. The current was switched by an arrangement of tappets outside the drill. He considered that, in the form originally conceived, the drill would never be successful as the tappet arrangement would cause sparking and be liable to considerable damage from exposure to the dirty and difficult conditions in mining.[18]

R.E.B. Crompton thought the first electrically powered coal cutting machine used commercially was installed by Messrs. Smith, Beaucock and Tannett of Leeds about 1884.[19] The provisional patent for this machine was taken out in November 1881 and the full specification deposited in May 1882, the patentees being J.R. Bower, J.E.A. Pflaum and J. Tannett.

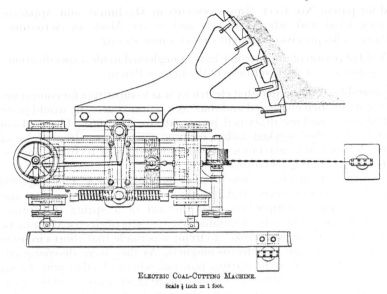

ELECTRIC COAL-CUTTING MACHINE.
Scale ½ inch = 1 foot.

Fig. 5.6 J.R. Bower, J.F.A. Pflaum and J. Tannett electrical coal cutting machine. Patent No. 5100.1881
Illustration taken from the *Proceedings of the ICE*, Vol. 104, 1890–91, p. 143

Like the machines previously described, this coal cutting machine operated with a reciprocating action (see Fig. 5.6) but the designers had moved away from the single pick to the multi-bladed cutter. The inference was that more power was now available and could be conducted down to the coal face.

The machine consists of two powerful electro-magnets mounted upon a substantial wrought iron frame, the line of action of the magnets running parallel to the coal face. The magnets were set opposite each other with a space between their faces in which was placed an armature connected to a rocker arm and thence to the cutting arm via a pivot. On the end of the cutting arm was fixed a quadrant-shaped head holding a series of picks, each set at an increasing radius from the leading pick.

The patent states that power was to be supplied to the machine from a suitable dynamo via a commutator or switching arrangement that fed current to each electro-magnet alternately. The commutator would be geared from the dynamo. Thus, with power applied first to one electro-magnet then the other, the armature interposed between them would be attracted to each in turn, producing a reciprocating motion on the quadrant arm. On each cutting stroke, a ratchet wheel was made to turn by the movement on the pivot and the whole machine travelled forwards on rails provided along the coal face by pulling on a short linked chain wound on to a drum on the machine. The other end of the chain was fixed to a metal support jacked between the floor and roof in front of the face. Further jacks supported the side rail which took the sideways thrust of the machine as the cutting head bit into the coal face.

This machine would be very awkward to handle and set up and would require the full depth of the cutter to be removed by hand before the machine could start on a new face. These and other problems could be alleviated by having a rotating drive mechanism rather than a reciprocating one, so three men from the Allerton Main Colliery set out to design a rotating bar cutter. Messrs. T. and R.W. Bower, proprietors of the colliery, and J. Blackburn, colliery manager, took out a patent in 1881.

During site trials of their rope-driven experimental machine, patent No. 10,120:1885,. Blackburn and Bower found great difficulty in providing the necessary power. Moreover, the driving rope connected to the main haulage system often came adrift from the pulleys and was pulled along the coal face at high speed, taking out props and threatening death or serious injury to any man in its way.

The problem of providing an appropriate drive for a cutting machine was emphasised by Blackburn at a meeting of the South Staffordshire and East Worcestershire Institute of Mining Engineers in 1885. Describing the new rope-driven machine, he said:

'Coal cutting machinery is a subject we have considered for the last sixteen years but, the greatest difficulty connected with this has been to secure a suitable motive power.'[20]

Having heard a little about the various applications of electric motors, Blackburn approached Fred Mori, amateur 'electrician'[21] who, amongst other things, had made fairground electro-mechanical contrivances and electric shock machines, and had once worked in some capacity under Faraday. He, in turn, approached L.B. Atkinson. Atkinson's account[22] of how he and his company became involved in helping to produce the first electric motor driven coal cutter suggests that the world of the early entrepreneurial electrical pioneers, whilst challenging and at times dangerous, had its more curious aspects.

Mori initially refused to reveal why he wanted a motor or what it was to be used for. Nevertheless, he was supplied with a 10 hp motor complete with full operating instructions. Some time later, Messrs. Goolden and Trotter were taken into the confidence of Messrs. Blackburn and Bower. L.B. Atkinson was then able to collaborate fully on a properly co-ordinated design and spent many hours at the coal face developing the coal cutting machine.

Fig. 5.7, taken from the drawing that formed part of Blackburn and Bowers patent No. 10,120:1885, shows the original design. It was a relatively simple layout with the rope drive turning a central shaft on which was mounted a crown wheel. This in turn drove the pinion wheel which was connected directly to the cutting bar. The platform on which the cutting bar and pinion wheel assembly was mounted could rotate, its movement being controlled by the long adjustable screw *t*. Thus, when the machine was presented to a new coal face, the cutting bar would be positioned as shown in the plan view of Fig. 5.7 and, as the adjustable screw *t* turned, the cutter would bite into the face. Once the bar was at right angles to the coal face, the carriage would be pulled along the face by hand.

In Atkinson's initial modified design, the motor was linked to the cutting bar through a chain produced by Hans Renold, who had recently established his business in Manchester, after patenting his bush roller chain in 1880[23]. Either because of over-ambitious expectations of the chain[24] or poor rigidity of the frame or motor mountings, this arrangement was not a success: the chain

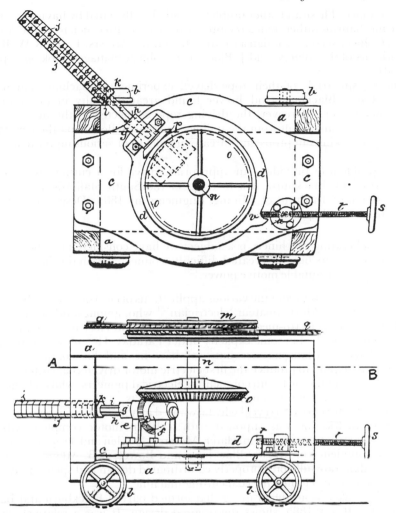

Fig. 5.7 T. and R.W. Bower and J. Blackburn's rope driven coal cutting machine.
Patent No. 10,120:1885
Illustration taken from the patent

repeatedly fouled the gearing and broke. The chain drive was then abandoned in favour of double helical gearing.

At this time, it was common practice for armature windings simply to be wound upon the smooth surface of the armature, which meant they could sometimes slip. In this instance, the vibration from the cutting bar caused the winding to be constantly shifting on the core, until the insulation was cut through. This forced the designers to adopt notched armatures to protect the coils.

This prototype, patent number 1857:1887, proved a very useful testbed, where a number of fundamental problems were identified. In spite of regular

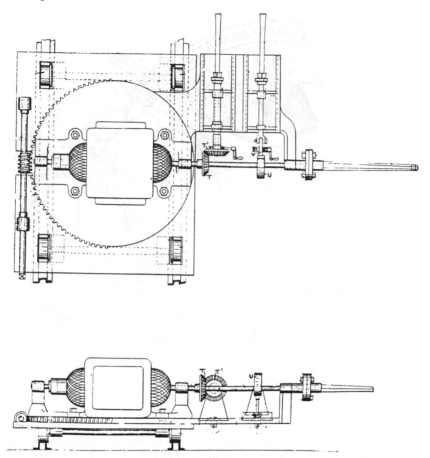

Fig. 5.8   T. and R. Bower, J. Blackburn and F. Mori's coal cutting and drilling machine.
Patent No. 7662:1887
Illustration taken from the patent

breakdowns and the prohibitive cost of repairs, good experimental results were
obtained and both the manufacturer and colliery management were encouraged
to continue to evolve better machines. A total of three patents were taken out in
1887 as designs were developed. The second of these, No. 7662, is shown at Fig.
5.8. The main drive gearing was here dispensed with and the cutter bar
connected directly to the motor shaft. In addition, the patent describes two
methods of drilling or boring that the machine is able to carry out: with one drill,
circular motion is provided by bevel gears T and T′, or with a percussive drill, the
'tamping' is provided through the cam U.

With the third patent, No. 16,955:1887, the drilling attachments have been
removed along with the name of F. Mori.

A feature retained in both improved versions is the semi-rotating platform,
allowing the cutting bar to be angled when starting a new cut. Lowering the

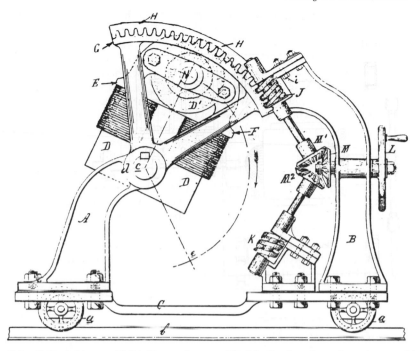

Fig. 5.9 L.B. Atkinson, H.W. Ravenshaw and F. Mori's electric coal cutting machine. Patent No. 14,090.1887
Illustration taken from the patent application

cutting bar to a foot or so above floor level would allow the undercutting to be done nearer the bottom of the seam. If the floor was soft, the track could be excavated to allow the cut to be made in the dirt, avoiding waste of coal.

The motor fitted to the modified machines was by Immisch & Co. of London and is of the open type. Neither patent mentions protection for the motor, but some form of covering must have been provided.

It is not clear why Bower & Blackburn abandoned the Goolden & Trotter motor in favour of the Immisch machine, but it may be significant that some two months before this latest patent was filed, Goolden & Trotter had entered the field with a coal cutting machine of their own, patent No. 14090 taken out in October 1887. The patentees were L.B. Atkinson and H.W. Ravenshaw, both designers with Goolden & Trotter, and Mr. F. Mori, who appears to have changed camps.

As can be seen from Fig. 5.9, the Goolden & Trotter machine was much more robust but possibly more cumbersome than the Blackburn & Bower cutter. The principal feature of the Goolden & Trotter machine was the ease with which the cutting height could be adjusted. This would be an advantage in cutting dirt bands found at varying heights in certain coal seams. The machine was also designed to pull itself along the coal face during the cutting operation. The ratchet and pawl controlling this forward movement was connected to an electro-magnet which automatically disengaged the feed if the motor was subjected to overload. This would be likely if the cutter bar got jammed or if a hard section of mineral, such as ironstone, was struck.

A feature similar to that on the Blackburn & Bower machine was the ability to carry out drilling work. This was not very well described in the patent and the drawing provided did not attempt to show how the mechanism would operate. This addition was evidently something of an afterthought.

As with the Blackburn & Bower machine, no mention is made of motor protection, but leaving the armature, commutator and brushes exposed, even without the complications of pit gas, would have given problems from all the inevitable grit and dust. Enclosing the motor would immediately present the risk of overheating and, as motor designers, they were fully aware of this.

In 1888, a further patent was taken out by L.B. Atkinson and F. Mori, now with the assistance of Fred Walker of Leeds. This machine was designed for driving headings, tunnelling or sinking shafts and could be fitted with either one or two drilling/cutting bars, as shown at Fig. 5.10. They have clearly moved away from the open type of motor (although the type of motor is not stated), and some form of protection appears to be intended, perhaps along the lines of the patent No. 12,676:1887 shown at Fig. 4.2 or Fig. 4.3

Goolden & Co. used their earlier 'flameproof' motor in a much improved

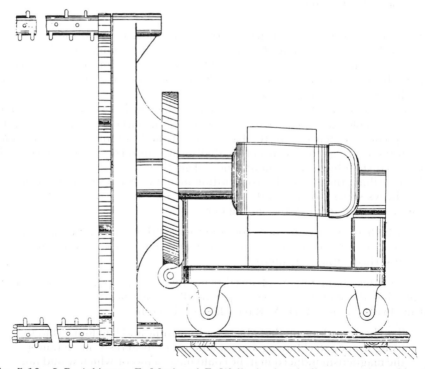

Fig. 5.10   L.B. Atkinson, F. Mori and F. Walker's electrically powered machine for tunnelling or shaft sinking. Patent No. 17,043:1888
Illustration taken from patent

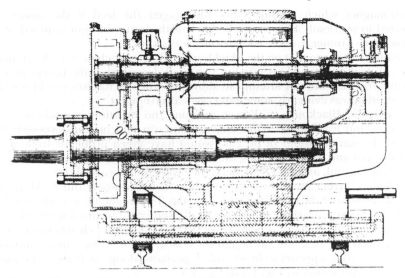

Fig. 5.11 I. B. and C.W. Atkinson and W.T. Goolden's electric coal cutting machine. Patent No. 14,041:1889
Illustration taken from the patent

design of coal cutting machine, patent No. 14,014:1889, an illustration of which is shown at Fig. 5.11 The motor for the machine was of the horizontal double field type and the patent states that the field windings were protected by steel or copper sheeting. The armature was totally enclosed and protected in accordance with their patent No. 12,676:1887.

The bottom half of the machine was made from a solid casting and contained the bottom pole pieces, cutter bar bearings and the circular base on which the cutter pivoted. Extensions to the base block at the front and rear of the motor carried the bearings for the armature shaft, which in turn was connected to the cutting bar shaft by means of double helical gearing.

In describing this particular coal cutting machine, Snell says that the motors were always series wound, so the speed of the motor varied inversely with the resistance encountered by the cutting bar:

'which is exactly the best condition for work of this kind, since it prevents undue shocks to the motor and tends to keep the power absorbed within safe limits'[26]

However, this characteristic of speed being inversely proportional to load also means that the motor will race to potentially dangerously high speeds if the cutting load is removed. H.W. Ravenshaw said that he:

'had been working on the subject for some time and had found that by using iron or steel having a great amount of hysteresis, i.e. retaining the magnetism induced in it, he could make a motor which would not race.'[27]

He explained that, by using the results of Dr. Hopkinson's work in this field, he was able to control the motor speed between 600 rpm on full load and 780 rpm on light load. To achieve this, he had used steel magnets and split the field windings into two circuits, arranging through an automatic switch to connect them in parallel when working at full load and in series when working below half load.

The patent, however, was evidently for the unmodified machine, as it was stated that 'the magnetising coils are wound on wrought iron cores' but there was no direct reference of the material of the yoke or pole pieces. Fig. 5.12 is a copy of the patent drawing showing the starter switch and wiring arrangements, this indicates a simple series motor not incorporating Ravenshaw's system of speed control.

With the switching arrangement shown, it would be reasonable to assume that the moving contacts were of the 'make before break' type, reducing the tendency to arc. Starting from rest, the contact arm would be on stud b and as it moved slowly from a_1 to a_4, sections of resistance would be shorted out and the motor would run up to speed. When stopping the machine, the reverse action would be taken. This switching arrangement will work without the cable link between stud b and the motor winding, but this connection, as explained in the patent, provides a shunt circuit through which the stored energy in the collapsing field of the machine (as the machine is brought to rest) may be safely dissipated. This prevents a sudden discharge and the inevitable arcing at the contacts.

This starting switch, which was mounted on a bracket fixed to one of the field winding cores, was 'practically air and gas-tight'. For enclosures where little or no heat was generated, this was an acceptable arrangement, for cooling air would not be required nor would the air inside tend to expand and contract in the cooling cycle drawing in a gas and air mixture.

While designing their later coal cutting machines, C.W. Atkinson said[28] that they found it necessary to look closely at the action of the cutting bar. When he and his brother entered this field, they had used the Blackburn cutting bar which was intended to cut the whole of the coal and had the cutting teeth one after the other on a six-tooth star wheel, each wheel sliding on the bar separately. They soon came to the conclusion that this was wrong and it was better to imitate as far as possible the action of the miner's pick. This they did by placing the teeth in line down and round the bar to form a single threaded screw with a pitch of one inch. Thus each tooth cut a narrow groove about ⅛ inch deep and the edge split the coal off into the next, so the main action was by splitting, not cutting. This produced less fine dust and, more importantly, reduced the power required by at least half, thus producing a load characteristic for which the early dc machine, with its limited output, was better suited.

A further improvement mentioned by L.B. Atkinson[29] was to fix the separate cutting teeth on the bar by driving them into tapered holes. This system had been invented by Hans Renold, patent No. 13,369:1889, and appears to have been incorporated into the Goolden cutting bar.

The cutting bar on this machine would, of course, need to be well supported and substantial journal bearings were provided, as can be seen at Fig. 5.11. The armature shaft was supported on two sets of journals, the lubricating ports being clearly shown. Communication channels took oil from the front port to lubricate the cutter bar bearings and drive gears, which ran partially submersed.[30]

An improved version of this machine was patented by L.B. and C.W. Atkinson

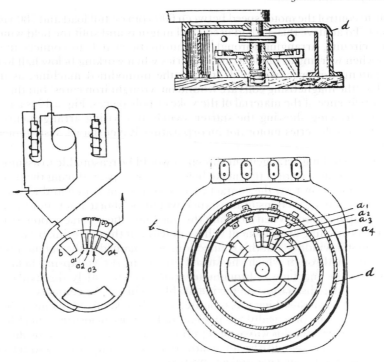

Fig. 5.12 L.B. and C.W. Atkinson and W.T. Goolden's electric coal cutting machine. Patent No. 14,041:1889. Details of electrical control
Illustration taken from the patent application

in 1890; patent No. 19,039. The motor of this machine was described as being protected by 'cases and jackets throughout, preferably according to the arrangements described in Patent No. 1, 12,676:1887, 16,623:1886, 536:1891'. The armature drive shaft transmitted power to the cutter bar via bevel pinion gears. This later machine also had the facility to allow the cutter bar to be swung round, into and out of the cutting position, by hand. There was also a special feature that allowed the bar to cut at rail (i.e. floor) level but would lift the bar clear of the rails as it swung away from the coal face.

A further, more novel, facility was that the machine could be split into two separate sections; the cutter bar and gear box in one section and the motor and drive shaft in the other. This arrangement allowed the motor to be removed 'from one machine of specific description, such as for right-hand cutting, to be attached to another, say for left-hand cutting'.

With a few notable exceptions, as stated earlier, there was a gradual move in favour of longwall working in most districts of the British coal field towards the end of the nineteenth century and into the twentieth. During this period, the disc type of coal cutting machine appears to have been the most popular choice of the small band of colliery owners and managers willing to experiment with new technology. Recognising this potential market, the Atkinson brothers took out a patent for such a machine in 1891. Their machine attempted to answer a number of the problems associated with coal cutting machines in general and disc cutters in particular.

One of the main problems with all disc cutters arises when, in cutting large sections of coal, some drops onto the disc and jams it. This was most likely to happen with discs supported on a centre bracket but otherwise left open. In the Atkinson design, the disc was supported and covered by two plates (one above and one below) which in turn formed part of the disc support bracket. The only part of the disc left exposed was the rim carrying the cutting teeth. On the inner edge of the rim, recesses were formed to make a near dust-tight fit with the protection plates and these were supported on a central journal and 'antifriction rollers'. Power was supplied to the disc via a pinion wheel that engaged with teeth set into the inner edge of the cutting disc rim. All the gearing was fully covered, thus protecting the operator and reducing the risk of fouling from coal dust.

It is obvious that the disc type of cutter would operate best on longwall faces with flat floors but would have difficulties with undulating floors. In an attempt to get round this, the Atkinson patent allowed for the drive motor to be supported on a separate carriage, with the two drive shafts connected via a universal joint. It was claimed that this division would help where it was 'necessary or advisable to reduce the weights of individual parts of the machinery'. Again, the patent gives no details about the motor or control gear.

Thus, by the start of the last decade of the nineteenth century, W.T. Goolden & Co. had produced a small range of electrically powered coal cutting machines and, in an effort to exploit these, a new company was set up, The Electrical Coal Cutting Contract Corporation, under the control and management of Claude W. Atkinson. This company took out contracts with coal owners to cut the coal and carry out all the necessary oversight and supervision; the coal owners being required to remove the coal and provide sufficient labour to work under the management of the Contract Corporation.

Initially, the contract system appeared to work well, but many coal owners were of the opinion that the system worked too well in favour of the Corporation. They wished to have a greater share in the profits. To this end, many of the coal owners bought the plant and operated it themselves. However, few of the owners fully understood the implications of opting to manage men and machines and consequently the system broke down.

The key man in the Contract Corporation's system of working was the underground organiser and supervisor. His job was to ensure that the work was well planned, with stoppages kept to a minimum, thus assuring a continuous production of coal. His salary was high — usually higher than the regular underground manager. To increase their immediate profit, the coal owners usually dispensed with the supervisor's services, following which failure of the scheme seemed inevitable.[31]

The newly formed Electrical Coal Cutting Contract Corporation found it far from easy to build up their business, although in a number of cases they were invited to recommence their operations, following the failure of the owner-managed scheme.

In 1893, W.T. Goolden & Co. merged to form Easton, Anderson & Goolden. The comments of L.B. Atkinson, a director of the new company, aptly sum up the position with regard to coal cutting machinery in Britain at that time:

'This company found a wide outlet for electrical machinery in directions occasioning much less anxiety and strain and therefore practically dropped the operation of the coal cutting machine. Frankly, the coal cutter was too far ahead of its time. The coal trade was too prosperous to think about saving the last shilling and the management was organised on too individualist a basis to make machine working effective. Thus, closed the first chapter in the history of coal cutting.'[32]

This is clearly a bitter comment, expressing the sense of lost opportunities, and was made in 1923 with some thirty years hindsight by a man then at the pinnacle of his career. By then L.B. Atkinson had been a Director of a number of important British companies and associations and President of the IEE 1920–21.

There may well have been a lack of enthusiasm amongst the colliery owners and managers to risk capital in this new field but the same cannot be said about design engineers and inventors. Between 1890 and 1900, a large number of patents were taken out in Britain, for new or modified forms of undercutting machines. The majority of these were not specifically designed to be driven by electric motors but many suggested electricity as an alternative power source. The patentees were thus aware of the viability of electric motor drives but perhaps lacked competence or confidence to apply them to their particular machines.

In 1892, J.B. Simpson, President of the North of England Institute of Mining & Mechanical Engineers, commented:

'The substitution of machinery for hand labour in hewing coal, introduced many years ago, does not seem to have made much progress.'[33]

By comparison, the picture in the USA was a little more optimistic. In 1891 there were some 545 coal cutting machines in operation, producing some 6.66% of that country's output of bituminous coal. The majority of these machines, of which the most popular appears to have been the Jeffrey parallel bar or face cutter, were powered by compressed air. This machine had a bar some 39 to 42 inches in length and the depth of cut was usually about 5 feet. It overcame the problems of working within the pillar and stall system (then the principal method of working in the USA) and undulating floors.[34]

In a paper to the American Institute of Mining Engineers in September 1890, J.S. Doe reported that his employers, the Jeffrey Manufacturing Co., had either installed or had orders for 23 of these machines powered by electricity. When first produced about 1879 the machine was powered by compressed air. The first trials of the electrically powered units were carried out by the Shawnee and Iron Point Coal & Iron Company mine in Perry County, Ohio in April 1889. The machines then installed were rated at 15 hp at 220 volts and were little different from those used at the original trial.[35]

W.S. Gresley described an interesting installation to members of the Institution of Civil Engineers in 1897.[36] His paper gave details of a central electricity generating station feeding five drift mines in Pennsylvania, USA. At two of these mines, a total of thirteen of the Jeffrey bar cutters were in use, working seams that sometimes gave off explosive gas and were sometimes wet. The paper also referred to another mine in the district where three chain cutters were in use, each rated at 20 hp connected to a 550 V 3 phase supply, the plant having been installed by GEC.[37]

Fig. 5.13   Jeffrey electrical parallel bar cutter *c.* 1893
Illustration taken from *Electric motive power* by A.T. Snell

In December 1893, trials were started at the Cannock & Rugeley collieries in the Midlands with two electrically powered Jeffrey bar cutters, one designed to cut a slot 4 feet deep and 3 feet wide, the other a slot 6½ feet deep and 3½ feet wide. In reporting the results of these tests, Robert S. Williamson spoke of machines used on the 5 feet thick bass coal seam where holing was very hard. His comments on the performance of the machines was not very complimentary. Having admitted that the general construction of the machines was satisfactory, he expressed the view that the motor lacked power and the machines were too light, though he appreciated they had been designed this way for ease of moving. He thought that the machines were better adapted for seams with good roofs and for non-fiery mines because of the open nature of the commutator and controls. A further drawback was the length required from the working face to the back timber or cog. If timber had to be taken out and reset, it would involve considerable danger, greatly impeding the work and adding to operating costs. During the trials, they had found that the machines were subject to a lot of wear and tear. He thought the wearing parts should be made stronger and that it would be very desirable to have spare parts available that were made in England.[38]

Table 5.1 Coal cutting machines used in British and American pits 1903–18

| | United Kingdom | | | | | | | | United States | |
| | Number of each kind of machine | | | | | | Total output of coal obtained by machines | Output per machine | No. of machines | Bituminous coal mined Statute tons |
Year	Disc	Bar	Chain	Percussive	Rotary heading	Total	Tons	Tons		
1903	483	39	25	77	17	643*	5,245,578	8,158	6,658	69,620,441
1904	516	60	34	122	23	755	5,744,044	7,068	7,671	70,261,158
1905	580	103	51	188	24	946	8,102,197	8,565	9,184	92,318,261
1906	656	152	37	254	37	1,136	10,202,506	8,981	10,212	106,113,863
1907	802	227	74	364	26	1,493	12,877,244	8,625	11,144	123,703,413
1908	827	261	94	451	26	1,659	13,508,510	8,143	11,569	109,985,120
1909	847	276	100	448	20	1,691	13,728,902	8,119	13,049	127,229,355
1910	930	333	134	556	6	1,959	15,747,558	8,039	13,254	155,368,119
1911	796	391	371	579	9	2,146	18,309,269	8,532	13,819	159,068,961
1912	1,103	468	182	680	11	2,444	20,053,082	8,205	15,928	187,981,091
1913	1,243	542	250	841	21	2,897	24,369,516	8,412	16,379	216,447,958
1914	1,262	585	294	926	26	3,093	23,976,367	7,752	16,507	194,999,363
1915	1,224	553	383	908	21	3,089	24,087,684	7,798	15,692	217,176,385
1916	1,255	569	520	1,095	20	3,459	26,303,110	7,601	16,197	253,295,960
1917	1,241	606	722	1,209	21	3,79C	27,626,298	7,272	Not available	Not available
1918	1,259	695	791	1,373	23	4,141	27,322,980	6,761	Not available	Not available

* Including two machines, the kind of which is not stated.

If Williamson's findings were typical of the Jeffrey machine operating in the British coal field at that time, the American company would need to make substantial improvements if they wished to increase their British sales. Some indication of the improvement made over the next 18 years is shown by British patent No. 8895, taken out by Jeffrey & Co. in 1911. This machine, designed for longwall working and cutting at the base of the seam, appears to be solid and compact. The patent states that the principal object of this improvement was to make the machine quieter in operation and give the miner a better chance of hearing the roof cracking, giving possible warning of a fall or total collapse. This quieter operation is achieved by the use of worm drives and spur gears to drive the cutting chain and self hauling gear, respectively. Earlier machines evidently had noisier spurs and bevel gears. The drive shaft extended from both ends of the dc motor, each end being fitted with worm gearing. One end meshed with gears that drove the chain cutter, whilst the other drove a set of reduction gears that propelled the machine along metal rails which ran parallel to the coal face.

By 1914 a total of 3,093 coal cutting machines were being used in the British coal field. The breakdown of these is shown at Table 5.1. From this Table, it will be seen that a significant number of percussive machines were in use, particularly in the Newcastle and Manchester districts. The overwhelming majority of these were compressed air powered, with only four powered by electricity.

Percy Fox-Allin states[39] that at the beginning of the twentieth century, there were only two types available, both American: the Ingersoll–Sergeant which was quite popular and the Harrison which never got beyond the development stage. The introduction of the 'Siskol' coal cutter in 1901 led to the development and increased use of percussive machines. It was the first to be worked from a column, could stand up to the hardest coal and could be used for undercutting or shearing from any angle. The application of electric drives to percussive coal cutters and drills appears to have presented designers with serious difficulties, although from about 1890 onwards, quite a few British patents were taken out. The first of the electrically powered tools were all solenoid operated but in 1894, Jeffrey & Co. applied an electric motor to a rock drill under patent No. 1914.

5.4 The gradual increased use of coal cutting machines in Britain

By the end of the nineteenth century, there was in Britain a slow but steadily growing realisation that coal cutting machines offered advantages over traditional hand hewing techniques. In 1900 the Government, wishing to obtain accurate data on the number of machines being used and the amount of coal produced, instructed all HM Inspectors to provide full details in their annual returns. The picture that emerged was less than encouraging. In March 1902, after two years' figures had been made available, the Home Secretary, the Right Hon. C.T. Ritchie, voiced his disappointment:

> 'The one great complaint I have to make against the mine owners is that they are not availing themselves of coal cutting machinery to anything like the same extent as mineowners in the United States.'

Between January and June of 1901 the *Colliery Guardian* ran a series of articles on coal cutting machines, giving comprehensive details of the various types of machine available, listing their strengths and weaknesses and explaining how the machine could be used to the best advantage. In 1902, this series was brought together in book form by S.F. Walker as one of the *Colliery Guardians'* series of handbooks under the title *Coal cutting by machinery in the United Kingdom*. In what appears to be an effort to emphasise Britain's poor showing in this field, he quotes the Home Secretary's remarks on the title page.

W.E. Garforth was a mining engineer and successively Agent, Managing Director and Chairman of Messrs. Pope & Pearson's Collieries, Normanton, Yorkshire, and founder of the Diamond Coal Cutter Company. He was knighted in 1914. In March 1902 he presented a paper to the Midlands Institute of Mining, Civil & Mechanical Engineers on 'The application of coal cutting machinery to deep mining'.[40] In this, he drew attention to the growth of international competition in supplying coal, the question of the duration of coal resources in the UK, and the need to guard against needless waste of coal. In linking these topics together, he advocated the use of machines to undercut the coal, citing the increased use of machines in the USA, and stated that in most circumstances machines had numerous advantages over hand working. These included:

- More coal could be extracted per man.
- The coal could be in better condition with less small coal produced.
- Cost per ton of coal produced was significantly reduced.
- The workmen's wages were higher.
- The accident rate was reduced and hence compensation payments were lower.
- Seams that were too thin or had hard coal, could be profitably worked that previously would have been abandoned.

In September 1902 Garforth provided further detailed evidence to substantiate his views to the Royal Commission on Coal Supplies.[41] The Commission had been given the extensive remit of reporting on:

- The resources of the British coalfield.
- Their probable duration.
- Possible economies.
- The effect of export of coal on British consumers and the Royal Navy.
- Maintenance under existing conditions of the competitive power of the British coal mining industry with the coal fields of other countries.

In general, Garforth's views were upheld by a number of other witnesses called to comment on machine working or other ways of making economies. These included:

- A.S.E. Ackerman, consulting engineer with special knowledge of American mining methods and machines.
- E. Bainbridge, mining engineer and colliery proprietor.
- G.E. Stringer of Stringer & Jagger, Park Mills and Emley Moor Collieries, Yorkshire.
- G.B. Walker, Chairman and Managing Director of Wharncliffe-Silkstone Collieries, Barnsley.

They also largely agreed with the opinions of colliery managers and others who had replied to the Commission's questionnaire on the advantages of coal cutting machines. These replies are contained in the third and final report dated 1905.[42] In summarising the various replies, many of which contained various qualifications and expectations, the report states that: 'The general tone of answers is very decidedly in favour of machines', and various individual comments are quoted, such as:[43]

- I am so satisfied with reduced costs that I intend gradually to abolish hand labour.'
- 'Nearly all the coal-beds in this country could be more safely and more economically worked by machine.'
- 'Generally I should hole coal by machinery everywhere if the seams were suitable.'

Asked if there were any disadvantages in working with coal cutting machines, Garforth replied that there were, mainly in the capital and maintenance costs of plant and equipment. He accepted that, in mines troubled with faults, the working could be more difficult. His calculations indicated that the capital cost could be redeemed in about ten years. From the comments of the Committee, this appeared to be an acceptable rate of return on capital.[44]

Both capital and operating costs were key features in the report on *Mechanical coal-cutting*, commissioned by the North of England Institute of Mining & Mechanical Engineers and published in 1905. This extensive report, running to 108 pages, was in two sections, Longwall Machines and Heading Machines.

The 21-man committee obtained written information from some 80 colliery companies, covering about 200 machines, and visited some 30 collieries. The costings relate to the closing months of 1902.

With longwall machines, savings compared to hand labour varied considerably, from over one shilling per ton down to no saving at all. When the figures were adjusted to take account of the cost of power, maintenance and interest on capital, the savings under average conditions amounted to about 4 pence to 6 pence per ton raised.

The committee reported that they were unable to obtain statistics showing savings or otherwise from the use of heading machines, but concluded that there were so many uses to which these machines could be applied with advantage and economy that their further introduction might be looked for.

However, as can be seen at Table 5.1,[45] there was a gradual increase in the number of rotary heading machines in use from 1903 to 1906, but then the numbers fell back and by 1918 they had returned to the 1904 level.

The three types of longwall machine considered were disc, bar and chain. In all three classes, electrically powered machines were found.

In the disc cutter class, three manufacturers produced electrically powered machines:

(i)   *Diamond*: this machine was unusual in having two motor drives, one at either end of the machine, which were fed from the 500V dc supply in series so that 250 V were dropped across each motor. This, the report said, reduced the risk of burning coils.[46]

(ii) *Jeffrey*: the report says that this machine had recently been introduced and was generally driven by electricity, though it was also made for compressed air. Operating voltage is quoted as 450–500 V.

(iii)   *Clarke and Stephenson*: this machine was one of the earliest to be electrically driven and is of the Gillott and Copley type. The motor's rating was 26 to 30 hp.

Two manufacturers made bar type machines:

(i)   *Hurd*: All the machines up to that time were electrically powered, though the report said that a compressed air model was in the course of construction. The machine was made in three sizes for cuts of 3½, 4½ and 6 feet. They weighed about 20, 30 and 45 cwt and were fitted with 12, 18 and 26 horsepower motors respectively. The voltages are quoted for two machines at 400 and 450 V dc respectively; the third machine inspected was three-phase 350 V.

The report confirmed that three-phase induction motors had been applied successfully to coal cutting machines, of both disc and bar type. Experience at Ackton Hall collieries, where they were first applied for this purpose, indicated that they were practically free from breakdowns. However, the problem of low starting torque was recognised and got round by allowing the motor to run up to speed before engaging the cutter bar or disc.

(ii)   *Goolden*: This type of machine was exclusively powered by electricity and the machine inspected was rated at 300 V dc. The machine was somewhat lighter than the Hurd machine and did not have the advantage of the Hurd reciprocating bar that assisted in clearing cut material. This machine was no longer being made for sale.

Only two chain type of machines were reported on:

(i)   *Morgan–Gardman*: Two separate installations of this machine were inspected, one in the Durham coalfield and the other in Northumberland. A peculiarity of this machine was that it ran on skids rather than wheels and,

Table 5.2   Coal cutting machines in British pits, 1902–09

Name of mining district and No.		1902		1904		1906		1909	
		E	CA	E	CA	E	CA	E	CA
Scotland	1 & 2	18	67	59	111	171	134	342	138
Newcastle	3	5	18	11	36	24	63	47	178
Durham	4	20	10	42	31	55	77	67	68
York & Lincs	5	35	94	64	101	94	158	123	135
Manchester and Ireland	6	9	14	11	35	8	45	14	111
Liverpool and North Wales	7	9	14	11	35	8	45	14	111
Midland	8	44	49	65	64	70	70	106	89
Stafford	9	14	8	11	11	17	13	25	28
Cardiff	10	–	–	–	9	1	13	6	21
Swansea	11	–	6	–	–	3	3	18	6
Southern	12	–	2	1	2	2	8	9	14

Data taken from HM Inspector of Mines reports for 1902 to 1909
Electrically powered machines are shown in columns E
Pneumatically powered machines are shown in columns CA

when hauling itself along the face, the haulage wire rope was set at a shallow angle to the face so that the machine pulled itself into the face. A long fender then ran in contact with the face and kept the cut at the correct depth. The weight was some 40 cwt, so this method of hauling must have taken a considerable amount of power, particularly on soft floors.

(ii)   *Mather & Platt:* The committee reported that this company was making endless chain machines that had already been tried out and would shortly be set to work in two or three collieries. The prototype would be made in two sizes, weigh some 45 cwt, undercut to a depth of 6 feet, and would be fitted with a 30 hp motor. The other would have a 15 hp motor, weigh 22 cwt, be capable of undercutting 4 feet 6 inches.

Over the next few years, there was a slow but steady increase in the number of coal cutting machines used in Britain. Tables 5.2 and 5.3 show that initially most

Table 5.3   Coal cutting machines and conveyors in use in British pits 1914

Divisions (and division number)	Disc EL	Disc CA	Bar EL	Bar CA	Chain EL	Chain CA	Percussive EL	Percussive CA	Rotary heading EL	Rotary heading CA	Conveyor
Scotland (1)	515	119	216	7	12	–	2	40	2	–	128
Northern (2)	56	81	33	15	63	8	2	451	–	11	68
Yorkshire and N. Midlands (3)	157	175	87	60	104	35	–	90	2	6	116
Lancs/N. Wales/ N. Midlands (4)	7	117	19	34	8	5	–	209	–	–	27
South Wales (5)	3	1	38	50	3	7	–	29	–	–	63
Midlands and Southern (6)	35	14	20	6	29	20	–	94	–	5	6
Totals	773	507	413	172	219	75	4	913	4	22	408

Electrically powered machines in column headed EL
Pneumatically powered machines in column headed CA
Data taken from HM Inspector of Mines Reports for 1914

machines, both compressed air and electrically powered, were installed in the Midlands area (Districts 5 and 8) and two Scottish Districts. From about 1906 onwards, the Scottish pits took the lead in the application of coal cutting machines.

5.5 Introduction of coal face conveyors

The last major facet of underground coal production to be mechanised was that of placing the newly cut coal into containers for transport.

By the end of the nineteenth century, mechanised haulage systems using rope-hauled wagons were quite common. These ran along the main roads, with feeder systems leading from the coal face if the seams were of sufficient thickness to make this economic. In such cases, the wagons would run parallel and close to the longwall face and coal was loaded into them by hand.

F.J.H. Lascelles of Pickering in Yorkshire designed a combined coal cutting and loading machine (patent No. 18,516:1897). It could be powered by either electricity or compressed air. It had four disc cutting wheels, three to cut horizontally in the bottom, middle and top of the seam and one to cut at right angles to the others, parallel to the longwall face, at a depth equal to that of the horizontal cutters, so that two rectangular sections of coal were cut from the face. The coal was then moved along a track by a scraper chain, and at a suitable distance was deflected via a chute into wagons moving at the same pace as the cutting machine. This 'cutter-loader' does not appear to have been widely applied, if at all, but the basic idea was later successfully exploited by a number of companies.[47]

Fig. 5.14 The Blackett face conveyor at work in 2½ foot thick seam, Kibblesworth Colliery, Co. Durham *c.* 1904. View of end of conveyor delivering coal into waiting tub

The first successful conveyor appears to have been the Blackett Conveyor, introduced into the British coal field in 1902. This was developed jointly with Clarence R. Cleghorn, an American engineer. Details were presented in a paper to the Institution of Mining Engineers in 1905.

The Blackett conveyor consisted of sections of steel trough 6 feet in length, 10 inches high and 19 inches wide, mounted on an angle iron framework. Those sections were made up to the required length and an endless chain was made to run in the trough. The chain had shaped links to convey the coal. Beneath the trough the chain ran on suitable runners. A drive motor, either powered by electricity or compressed air, was sited at the delivery end of the conveyor, over which the coal dropped into waiting wagons (see Fig. 5.14).

In his paper, Mr. Blackett explained that the system was particularly suitable for use in thin seams, where it was not economic to run wagons down the length of the longwall face, as this involved removing substantial sections of floor or roof to accommodate the track and wagon.

As can be seen from Table 5.3, by 1914, 408 conveyors were in use in British pits. Most were in Scotland and Yorkshire, where coal cutting machines were more readily accepted and by then were well established.

5.6 Notes and references

1 MITCHELL, B.R.: '*Economic development of the British Coal Industry 1800–1914*' p. 70
2 PAMELY, C.: '*Colliery managers' Handbook*', 1898, p. 242. Long wall working was by no means universally adopted during the latter half of the nineteenth century. In certain areas, notably Northumberland and Durham, whether for geological reasons or the lack of will to change from traditional ways of working, the pillar and stall system remained well rooted.
3 'Steel wire ropes in mining practice', Paper no. 2 in the proceedings of a conference at Leamington Spa in 1950, published in book form as *Wire ropes in mining*, Institute of Mining and Metallurgy, 1951
4 KIRSOPP, J.: '*Use of power in colliery working*', 1926, p. 238
5 MITCHELL, B.R.: op. cit. p. 75
6 BUXTON, N.K.: '*The economic development of the British Coal Industry*', 1978, p. 99
7 *Engineering*, 1873, Vol. 16, p. 282
8 WALKER, S.F.: '*Coal cutting in the United Kingdom*'. 1902, p. 28
9 *Transactions North of England Institute of Mining and Mechanical Engineers*, 1867–68, Vol. 17, Appendix No. 2
10 The illustration is taken from *M. & C. Machine Mining*, Vol. 1, p. 137. The date this machine was built or used is not given
11 *Transactions Institution of Mining Engineers*, 1901–02, Vol. 23, p. 330
12 '*Report of the Commissioners Appointed to Inquire into several matters relating to coal in the United Kingdom*', 1871, Vol. 1, p. 117
13 *Proceedings Institution of Mechanical Engineers*, 1872, p. 211
14 *Transactions Federated Institution of Mining Engineers*, 1889–1890, Vol. 1, p. 123
15 *ibid.*, p. 141
16 It is apparent that an error has been made in the patent in referring to coil B instead of coil G and L.B. Atkinson, when referring to this machine in article entitled 'The early history of electric coal-cutting' in *M. & C. Machine Mining*, 1923, Vol. 3, p. 53, quotes the wording of this section of the patent without correction. In this article, no suggestion is put forward as to how the machine worked nor is comment made on its efficiency

17 '*Electricity Supply in the United Kingdom*' Electricity Council, 1983, pp. 4 and 8
18 *Proceeding Institution of Civil Engineers*, 1890–91, Vol. 104, part 2,. p. 151
19 *ibid.*, p. 142
20 *Transactions South Staffordshire & East Worcestershire Institute of Mining Engineers*, 1885, Vol. 11, p. 81
21 Today, the term 'electrician' is used to describe a craftsman or tradesman. In the latter part of the nineteenth century and for some time into the twentieth century, electrical engineers from the most eminent downwards were often given this title
22 '*M. & C. Machine Mining*', Vol. 3, Mavor & Coulson Ltd. 1923, p. 54
23 TRIPP, B.H.: '*Renold chains*', 1956, p. 11
24 *ibid.*, p. 82
25 About this time, Mr. Trotter decided to leave the company. L.B. Atkinson was made a partner and the new firm moved to London, intending to design an entirely new machine. As part of their drawing staff, they took on Mr. F. Hurd, junior, who was soon to establish himself as one of the leaders in the field of coal cutting machine design. His father, F. Hurd, senior, had been involved with the design of coal cutting machines many years earlier
26 SNELL, A.T.: '*Electric Motive Power*' The Electrical Publishing Co., London, 1894, p. 348
27 *Proc. ICE*, 1890–91, Vol. 104, Part II, p. 164
28 *ibid.*, p. 165
29 '*M. & C. Machine mining*' Vol. 3, 1923, p. 79
30 This would lead to quieter running which would be of some benefit to the miner when he was listening for movement of the roof or support timbers
31 HURD, F.W.: 'Electrical coal cutting machines', *Trans. IME* 1903, Vol. 25, where it is claimed that 'machine cutting demands good organisation and close and persistent supervision', (p. 110) and 'many failures in coal cutting are attributable to a lack of proper supervision' (p. 234)
32 '*M. & C. Mining machinery*', Vol. 3, 1923, p. 101
33 *Trans. NEIMME*, 1891–92, Vol. 41, p. 178
34 *Proc. ICE*, 1890–91 Vol. 104, Part II, p. 178
35 *Transactions American Institute of Mining Engineers*, 1890–91, Vol. 19, p. 276
36 *Proc. ICE*, 1897–98, Vol. 131, Part I, p. 100
37 *ibid.*, p. 116
38 *Trans. FIME*, 1893–94, Vol. 7, p. 305
39 'Mechanised coal cutting and conveying', *The Iron & Coal Trades Review*, 1949, Vol. 159, p. 109
40 *Trans. IME*, 1901–02, Vol. 23, p. 312
41 *Royal Commission on Coal Supplies, 1st Report 1903*, Cd. 1724, Q2675–2867
42 *Royal Commission on Coal Supplies, 3rd Report 1905*, Cd. 2353, Part XI, Appendix 7, p. 44
43 *ibid.*, p. 47
44 *Royal Commission on Coal Supplies*, 1st Report 1903, Cd. 1724, Q2675–2867
45 *Sankey Commission Report*. Appendix 34, p. 64
46 At this time, it was quite common for the armature cores of coal cutting machines to burn out. One insurance company reported that, of those machines insured by them, 51.5% broke down per year; and of these 74% were due to failure of the armature. *Report of the Departmental Committee on the Use of Electricity in Mines* 1905, Cd. 1916, Q5946–5948
47 In 1943, Anderson Boyes introduced their Meco–Moore Cutter–Loader. This machine used two horizontal chain cutters and one vertical cutter. See CARVIL, J.L.: '*Fifty years of machine mining progress*', p. 62

Chapter 6
Winders and haulage locomotives

6.1 Electric winders

6.1.1 Background philosophy

During the early years of this century, colliery owners contemplating installing a new winder could choose between the well-proven steam winders and the more recently introduced electric versions.

Because of their high current consumption and cost, electric winders were normally only feasible as part of an overall electrification scheme. However, a number of points warranted careful consideration before a final decision was arrived at.[1]

- The number and application of steam engines already existing on site.
- The location of the colliery and whether it was the only undertaking owned by that company
- The possibility of a number of collieries belonging to one group having a central generating station
- The possibility of using an external supply from a Public Supply Authority (PSA)

There was no easy choice, as W.C. Mountain indicated when, in 1904, he encouraged colliery owners to look carefully into the matter.[2] Recognising the expense involved and the fact that electric winders, certainly as far as Britain was concerned, were still very much a novelty, he advised against the indiscriminate use of electricity in a pit. 'Every case', he said, 'will have to be considered on its merits'.

Cost apart, many collieries could not easily install electric winders; they either lacked adequate planning to make a major change or were thwarted by physical problems, often arising from their location. The latter applied especially in South Wales where the narrow valleys decreed a haphazard positioning of steam plant, and even a moderate degree of electrification could have necessitated major reconstruction. Under such circumstances a colliery intent on installing new winders would probably opt for a boiler of sufficient capacity to supply the winders and all site services. Later, mixed pressure turbo-generators offered a more economical alternative, but these had to be chosen with care or they could restrict further development.[3]

Even when geographic conditions were favourable, many companies owning only one or two collieries could not normally justify the expense of erecting their own power station, although most had some electrification using semi-portable boilers and generating sets. If close to a public supply line, it was worth considering bulk purchase of electricity. Some power companies offered electricity below cost to collieries in order to improve their load factors which were often poor. This is discussed further in chapter 8.

Only the larger colliery companies could afford to install electric generating plant of sufficient capacity to power winders. Although they could not always generate as cheaply and efficiently as a well run public supply undertaking, e.g. NESCo, even when using inferior coal (which they could not sell), it gave them a high degree of autonomy.

By 1900 winders could still be regarded as in the experimental stage, even though two distinct types had evolved — geared ac and dc with a Ward Leonard (or variation) control. Geared ac types (initially slip-ring induction motors with liquid controllers) were smaller than the Ward Leonard, easier to maintain, cheaper initially, and even allowing for rheostatic losses, more efficient. The advantages of the Ward Leonard were smoother control, and regulation over a wide range of speeds and braking.[4]

6.1.2 Winding techniques

In deep shaft winding the weight of the winding rope often exceeded that of the coal being raised. Consequently, very high torque was required at the beginning of the wind, followed by heavy braking at the end to bring the winder to a standstill, particularly when a plain-cylindrical drum was used (Fig. 6.1a).

A number of techniques were developed to 'smooth out' the load curve. Two of the most popular were the tailrope (used with a plain-cylindrical drum) and the profiled drum.

In the first method (see Fig. 6.1b) a tailrope was attached to the underside of each cage and either allowed to hang down the shaft or guided by a pulley (sheave) in the sump. By designing the weight of the tailrope to equal that of the winding rope, a balanced condition was achieved, ensuring that the only load raised by the engine was that of the coal. Although the engine had less load to lift, the mass to be accelerated and decelerated during the wind was increased.

The second method involved the use of a profiled drum. The conical drum was first introduced by John Smeaton around 1780[5], and developed into that mainstay of deep winding — the bi-cylindrical-conical drum. This drum had cylindrical sections of two different diameters joined by coned sections (see Fig. 6.1c and d).

At the start of the wind, the rope connected to the full cage was on a smaller diameter, and the empty cage rope on a larger diameter (Fig. 6.1c). As the wind progressed, the positions gradually reversed, ending as shown in Fig. 6.1d. Since the torques virtually cancelled each other out, the engine only had to do sufficient work to accelerate the moving masses and overcome frictional effects.

Although this method did not have the disadvantages of the tailrope, it was more expensive. Some savings in initial cost could be made, however, in the reduced size of the electric motor since lower peak currents were taken during the load cycle. This reduction can be clearly seen from the load curves (Figs. 6.2a and b), which were obtained from comparative tests carried out in 1912 on an electric winder in a colliery on the English north-east coast.[6]

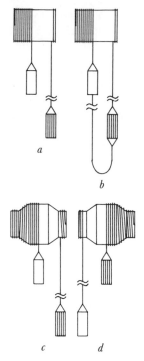

Fig. 6.1   Winding techniques
Based on illustrations in MASON, E: 'Practical Coal Mining', 1950, Vol. 2, p. 454

The Koepe system, introduced in 1877, overcame the high costs of cylindrical or conical drums. It utilised only one rope which passed over a single-grooved sheave instead of a drum, before being fastened to the cages via head sheaves. A tailrope connected to the bottom of the cages (around a sump sheave) ensured that the rope tensions were maintained at a constant ratio. The drive was entirely dependent upon the friction between the pulley and the rope.

The Koepe system became the standard method of winding in the Ruhr valley in Germany,[7] but its popularity was limited by its disadvantages: the rope was liable to slip on the drum, both cages would fall should the rope break, and difficulty in winding from different levels.

6.1.3 Hesitant first steps

Although electric winders were in use at the Lambton Colliery[8] and the Earl of Durham's Colliery[9] as early as 1891–92, they were small and their use was confined to staple winding (a shaft connecting two levels of underground workings).

In 1900 the Inspector of Mines for the Liverpool District stated in his annual report that the electric winder installation at the Broad Oak Colliery near St. Helen's was, to his knowledge, ' . . . the first instance of a motor having been utilised in this country for the actual winding in a coal shaft'.[10] This remark may well go unchallenged, because there is very little evidence to suggest that British

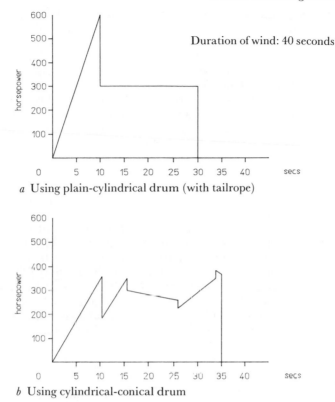

a Using plain-cylindrical drum (with tailrope)

b Using cylindrical-conical drum

Fig. 6.2   Horsepower/second diagrams for north-east coast colliery fitted with an electric winder

collieries showed any interest in electric winders, despite the activity on the Continent. One reason — and not necessarily the main one — was their reluctance to apply a relatively unproven technique when an established and economic one (steam) already existed.

The winder at Broad Oak was not new, but an electric motor was substituted for an existing steam winding engine. Both men and coal were raised up the shaft which was 61 yards deep.

Although the Mines' Inspector failed to give motor details (other than the shaft speed, 500 rpm, and the geared speed to the winding drum, 30 rpm), he mentioned that the voltage was 500 V and electricity was supplied from the St. Helen's and District Tramway Company at twopence per unit. He continued: ' . . . even at this somewhat high price there seems to be considerable economy as compared with steam.' It is a pity that this statement was not qualified because electricity was available from other power companies at a quarter of the price.

By 1904 an interesting system was working at the Shelton Iron, Steel & Coal Company Ltd., Brook Pit, Heckmondwike.[11] Although it was used for staple winding, it was introduced by courageous colliery owners willing to apply electricity to solve a problem on what appeared to be purely financial grounds.

A fault caused the coal seam being worked by the Brook Pit to drop vertically 180 feet from its original level. To continue working they needed access to the dip-side. Two techniques were available: driving a 1 : 6, 1,080 foot drift at a cost of about £1,960 or sinking a 180 foot staple at the cost of some £3.10 per foot. The winding gear subsequently installed raised 150–200 tons up the shaft in a ten-hour day, each wind taking 40–45 seconds.

The dc series-wound motor was capable of developing 40 hp at 500 rpm and had a number of special features, including a slotted-drum armature, large commutator with carbon block brushes and self-oiling bearings.

Motion was transmitted to the 3 feet diameter winding drum by a first and second motion gearing. A powerful foot-operated brake was fitted to the drum, and for additional safety there was also an electrically operated brake. This latter brake, similar to the type commonly used in Germany, was attached to the motor shaft and was of a fail-safe design (it clamped the shaft when the power was off).

The gamble taken by the colliery owners in choosing this system was certainly justified because they were able to win the coal in nine months less than they would have by the more conventional drift. They also saved £1,330.

The same year (1904) saw the installation of an Ilgner Electrical Winding Set (manufactured by Bruce Peebles of Edinburgh) at the Tarbrax Oil Co. Ltd., at Cobbinshaw, Lanarkshire.[12] It is possible that this was the first Ilgner system (one of the best methods of electrical winding yet introduced) to be installed in Britain. 'Firsts' are always difficult to substantiate, but there is little doubt that it was the first Peebles–Ilgner system to be used in a British colliery.

6.1.4 Continental developments

Most British engineers could see there were clear advantages in running small electric motors intermittently, or large motors continually, but were doubtful about their use where duty cycles varied considerably.[13] They were simply not convinced of the supremacy of motors over steam engines under such conditions. It is difficult to see why such attitudes persisted in view of the developments made by Siemens and AEG in Germany and other European countries. Ignorance was hardly an excuse because people like W.C. Mountain (of Ernest Scott & Mountain, Newcastle upon Tyne) went abroad to investigate these developments at first hand and disseminated their findings widely in papers to various learned societies.

In March 1904 Mountain visited Merklinde, Belgium, where one of the largest winders in existence was installed at the Zollern 11 colliery.[14] The winder, of the Koepe design, could raise forty 5 ton loads per hour from 330 m, which in a 16 hour day amounted to over 3,200 tons of coal. It was capable of far greater loads, for it was intended eventually to raise coal from the lower series 1,100 m deep.

The winder was driven by two Siemens & Halske motors, each capable of developing 705 hp at 500 V within the speed range 51–64 rpm. When the system was originally fitted, these motors were powered from two 1.1 MW generators, and connected across this supply was a 250 cell, 5,000 ampere-hours, 500 V battery to smooth out the peak loads. However, the complicated switching required to bring in various groups of cells proved unreliable, so a Siemens–Ilgner set was installed. Since the batteries were capable of running the

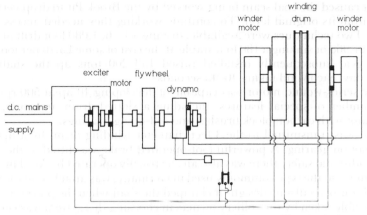

Fig. 6.3 Block diagram of the Zollern 11 Siemens–Ilgner winder

whole plant, including the winder, for one hour they were retained as a standby facility.

Fig. 6.3 shows a block diagram of the Siemens–Ilgner system used at Zollern 11.

Interposed between the incoming mains and the winder motors was a motor–generator (M–G) set capable of producing a full load current of 1,800 A at 550 V (on generator side). Connected to the shaft was a 45 ton, 12·5 ft diameter flywheel having a peripheral speed of 14,000 ft per minute. The mains-connected motor drove the flywheel and generator (connected on the same shaft). The generator current thus fed the two winder motors. Field current to the generator and winder motors was provided by a small generator, connected to the above shaft. Regulation — which determined the speed of the wind — was adjusted through rheostats fitted in series with the generator field winding.

When the excitation current was small, little or no current was supplied to the winder motors, although they could still be running on account of the inertia of the load. Under such conditions the generator received a reverse current flow, causing it to act as a motor, accelerate the flywheel and thus begin to store up energy.

At the commencement of a wind, load was put back onto the generator. The motor would then try to slow down, but this would be prevented by the flywheel, which now began to expend its energy and drive the generator to provide power to the winder motors, thus relieving the motor and supply system of excessive current demand.

Mountain inspected other colliery and salt-mine winders during his Continental visit.[15] Two collieries — the Arnim at Zwickau and Preussen 11, near Dortmund — had three-phase installations. Their sizes differed considerably: the former produced a maximum of 80 hp, while the latter had a 1,500 bhp motor directly coupled onto a Koepe drum and capable of winning coal up the 560 m shaft at 16 metres per second and men at 6 m per second. Problems were experienced at Preussen 11 because the voltage dropped sharply at the commencement of the wind.[16] Two 550 kW alternators had to be run continuously to prevent this voltage drop from affecting the remainder of the colliery drives. On a subsequent visit Mountain observed that the situation had not improved, when the winding engine was being used as a maintenance platform for men to fix cables in the shaft. Under such conditions the motor was doing little more than overcoming frictional losses; even so the system pressure had dropped from 2,300 V to 1,800 V, but admittedly only one generator was working. Despite the increased voltage drop, Mountain thought the ac winder more successful than the Ilgner system, but he still thought that dc motors were better suited for main winding gears.

Speed regulation and reversal of the slip-ring ac sets were achieved by liquid rheostatic controllers. In spite of losses, these were the most straightforward and economical method for drives up to 500 hp.[17] Liquid controllers of the lifting electrode or weir type were used, the latter being the most common because it permitted the setting of definite maximum acceleration. To improve conductivity within the liquid (water), 10% by weight of Carbonate of Soda was often added.[18] The reversing mechanism was also incorporated into the controller where a changeover switch simply reversed two of the three phases. Fig. 6.4 shows a typical weir type controller.

Control of the ac motor depended upon the movement of a lever linked to one arm of a three-arm crank. The water-filled tank was effectively divided into two: the upper or electrode chamber housed the fixed electrodes, which led to the rotor via slip-rings; the lower chamber contained cooling water. A circulating pump connected the two chambers, but the level of liquid in the upper chamber was regulated by a sluice, controlled by the lever. Movement of this lever also determined the acceleration time of the motor. Another advantage of this type of controller over the moving electrode type was that, in the event of an emergency, the throttle (or sluice) could be quickly opened and the water level reduced, thereby stopping the motor. In the moving electrode type, it would take several minutes for the worm gear to lift the electrodes clear.

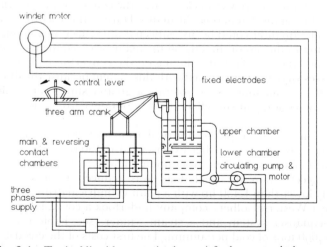

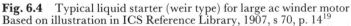

Fig. 6.4   Typical liquid starter (weir type) for large ac winder motor
Based on illustration in ICS Reference Library, 1907, s 70, p. 14[19]

6.1.5 British trends

The main drawback of electric winders, was the heavy, short duration loads that they drew at the beginning of each wind. Collieries often had winding duty cycles of less than a minute, so considerable problems were created by taking power from supplies of limited capacity, as at Preussen 11 Colliery. If a public supply was involved, in order to reduce inconvenience to other users of the system, equalisation was often insisted upon.[20] These systems were expensive and this, together with the higher no-load losses, deterred a number of collieries from taking advantage of bulk supplies.

Both types of winder were capable of equalisation to reduce these initial high currents. Although the Ward–Leonard took higher peak loads than its geared ac counterpart, they were of shorter duration. For this reason many of the early winders were Ward–Leonard types, incoporating some form of equalisation.

Equalisation of a Ward–Leonard system basically meant adding to the M–G set a heavy flywheel, which stored up or released energy depending on whether the system was running on no-load or winding. Whilst it was possible to achieve almost perfect equalisation by increasing the flywheel dimensions, practical limitations and cost restricted the peak load to about 15% above the mean demand.[21]

An interesting variation on the equalisation principle was that installed by Westinghouse at the Great Western Colliery Company's Maritime Colliery, South Wales, in 1908.[22] This system was not so much chosen as imposed by restrictions of speed, weight and depth of wind. In the convertor–equaliser system, the three-phase main was connected directly to the winder motor, in parallel with a transformer and rotary convertor. Under no-load conditions the current taken by the convertor drove a dc motor with the now familiar, heavy flywheel fixed to its shaft to store energy. At the beginning of the wind, the heavy load imposed on the motor decreased the current output of the rotary convertor, causing the flywheel to release its energy and drive the motor. The motor, now acting as a dynamo, fed power back through the convertor for the winder motor, as well as reducing the demand on the mains. During times of little or no winding, the convertor–equaliser could be disconnected thus effecting a saving. However, the facility to run the motor directly off the mains was retained — a useful safety feature in the event of a fault developing on the convertor–equaliser part of the circuit. Losses were further reduced with this system because the flywheel, being driven by the dc machine, could tolerate a greater slip — up to 30%.[23] This meant that the same equalising effect could be obtained from a flywheel of smaller dimensions.

The decision to adopt a spiro-cylindrical drum for this installation was a direct consequence of the success earlier experienced (*c.*1905) by the Powell Duffryn Steam Coal Company at their Aberaman colliery.[24] They found that during maximum acceleration the peak load was reduced to less than 1,000 hp, significantly lower than 1,450 hp with a plain-cylindrical drum.

The Great Western Colliery company undertook a cost analysis between the convertor–equaliser system and a steam winder (see Table 6.1)[25], based on raising 300,000 tons of coal per annum. The first costs of the electric winder was estimated at £110,000 compared to £12,000 for steam. Annual running costs for electricity were put at £2,887 16s. 3d. and for steam at £2,837 5s. 0d.: a total

Table 6.1 Comparison between first cost and annual cost of a steam and electrical winder, to raise 175 tons per hour, assumed to raise 800,000 tons per annum.

Steam Winder				Electrical Winder			
First Cost				*First Cost*			
High class twin compound steam winder, guaranteed to run 70 trips per hour on a steam consumption of 7,800 lb or 108 lb per wind, equivalent to 35½ lb per useful H.P.				Converter-equaliser three-phase winder, guaranteed to run 70 trips per hour on a consumption of 336 units, equivalent to 4.8 units per wind			
Engine-house and pillars, other than deep foundations, crane, sundry extras, and erection				Engine-house and pillars, other than deep foundations, crane, sundry extras, erection and testing.			
Boiler plant of 3 Lancashire 30' × 8½' boilers, bunkers, flues, seating, economiser, chimney, feed pumps, steam pipes, and covering, boiler shed, coal and ash plant and sundry labour				Share of cost of electrical sub-station and transmission line from same to engine-house Power to be purchased from Power Co.			
		£ s. d.				£ s. d.	
Complete plant as above, as per tenders and estimates		£12,000 0 0		Complete plant as above, as per tender and estimates		£11,000 0 0	
Annual Cost				*Annual Cost*			
		£ s. d.				£ s. d.	
Engine-room—				*Engine-room—*			
Wages, 3 enginemen and 1 cleaner		475 0 0		Wages, 3 enginemen and 1 cleaner		475 0 0	
Ropes, estimates life 2 years		80 0 0		Ropes, estimated life 2½ years		65 0 0	
Stores of all descriptions		50 0 0		Stores of all descriptions		50 0 0	
		605 0 0				590 0 0	

Steam supply—

	£	s.	d.
Wages, 8 stokers and 1 coal and cash man	335	8	0
Cleaning flues, 21 times @ 13s.	15	12	0
Preparing boilers for annual inspection, 3 @ 35s	5	5	0
Stripping boilers every 5 years, 3/5 @ £10	6	0	0
Firebars and sundry maintenance, 3 @ £10	30	0	0
Stores, viz. rings, glasses, oil, waste, &c., 3 @ £5	15	0	0
Maintenance and cleaning of economiser	50	0	0
Maintenance of pumps, pipes, and covering	50	0	0
Coal, 3,000 tons per annum @ 7s. 6d.	1,125	0	0
	1,632	5	0

Depreciation—

	£	s.	d.
5% on £12,000	600	0	0
	£2,837	5	0

On 300,000 tons per annum = 2.23*d.* per ton

Electrical supply—

		units
1,050 hours running		655,200 units
5,250 hours waiting		183,750 units
1,560 hours shut down		nil units
Total units per annum		838,950
838,950 units @ ½d	1,747 16 3	

Depreciation—

	£	s.	d.
5% on £11,000	550	0	0
	£2,887	16	3

On 300,000 tons per annum = 2.31*d.* per ton

saving of some £50 in favour of steam during the first year (this, related to the cost per ton of coal, gave 2·31d. and 2·23d. respectively). These differences were not of sufficient magnitude to prejudice owners against installing an electric winder.

The choice between the geared ac motor winder or the dc system with its Ward–Leonard control (or similar) remained open to the colliery owner. Both types, as have been shown, had advantages and disadvantages. Initially the high peak currents they produced were a nuisance. Several methods were developed to keep this to a minimum, but it seemed that the only ideal solution was to build a 'super-power station', whose capacity was such that these peaks would not be felt.[26]

In 1911 *The Electrician* published a comprehensive (though not exhaustive nor error free) list of electric winders installed in various parts of the world.[27] More than 400 installations were described, although not all related to coal-mining. The most noticeable feature was the dominance of German manufacturers in the British market (Siemens 52·3%, AEG 25%). Westinghouse, by comparison, accounted for 13·6% and Lahmeyer 9·1% and these are summarised in Table 6.2.

The number of Ilgner systems was predictably high, reflecting the now familiar problem of limited supply capacity, either self-imposed through financial constraints or determined by the supply authorities. Where capacity was not a problem the much simpler direct (or geared) winder was often chosen, especially now the 'super-power station' concept was being applied. This did not mean the demise of the dc types since local conditions, attitudes and costs and difficulties experienced in controlling the very large winders, particularly when banking, ensured their existence for many years to come.

Fig. 6.5   Electric winder installed at the Maerdy East Pit, Glamorgan, South Wales, showing a plain cylindrical drum

Table 6.2 Summary of electric winders installed in the British coalfield (1911)

Manufacturer	Ilgner	Ward Leonard	Direct coupled/geared	Convertor	Unspecified	Total
			Type of system			
Siemens	11	3	4	3	2	23
AEG	2		9			11
Westinghouse		1	4	1		6
Lahmeyer	2		2			4
Total	15	4	19	4	2	44

The distribution of electric winders throughout the coalfield was far from uniform: South Wales had the greatest concentration (52·3%), despite the earlier mentioned physical constraints. Deployment of the Ilgner and direct (or geared) types were almost equally divided. Most were installed in colleries which belonged to the large and often diversified companies. The Markham Steam Coal Company and the Tredegar Iron Company, which shared the same board of directors, had four; the Powell Duffryn Steam Coal Company, Limited, had seven, including the largest winders installed in Great Britain. These (see chapter 8) were installed at the Britannia Colliery, Pengam, and were capable of developing 4,350 hp. Equalisation restricted them to a more conservative, but still large, 1,750 hp.

Despite *The Electrician's*, optimism '. . . what great progress had been made in the application of the electric drive to winding plant', and the appearance of large, central generating stations, the amount of power consumed by winders in 1912 put the situation into more realistic perspective. Of a total of 510,756 hp installed in British collieries, only 23,896 hp (4·68%) was used by winders.[28]

Such figures did not prevent anybody from speaking without some degree of caution, for in that year A.E. du Pasquier said at Cardiff:[29]

> 'The progress made in electric winding has been so marked . . . that as time goes on — with the expansion of the present large power distribution companies, increasing cost of labour, increasing value of small coal, and the antiquation of existing steam winders and boilers — the steam winder will eventually become rather the exception than the rule as in the past.'

How he must later have regretted these remarks, for by 1937 the percentage horsepower taken by winders still remained in single figures — 7·66%.[30]

6.2   Underground locomotives

6.2.1   Development of the electric locomotive

One of the first collieries to adopt electric locomotives was the Royal Colliery at Zaukeroda, Germany,[31] in 1882, when ever increasing demands on the horses made them introduce mechanical haulage.

The working was 720 m in length and 220 m below the ground, with a fresh-air intake, so severe constraints were placed upon the type of system to be used. Stationary steam engines for rope or chain drives was not considered desirable and compressed air virtually excluded itself because of its inefficiency. This left electricity as the only viable alternative. The decision was made a little easier since the colliery had earlier installed an electrically operated underground ventilator with good results.

Siemens & Halske were asked to supply a locomotive capable of drawing a gross load of 7·5 tons at a speed of 2 m/s in either direction. This was no mean task because scarcely three years had elapsed since they first successfully demonstrated an electrically powered vehicle at the Berlin Exhibition. That vehicle could hardly have been regarded as suitable for industrial purposes.

Power to the locomotive was provided by a dynamo (Siemens No. D 0 size) located at the surface. Although bare conductors ran from the dynamo to the pit-bank, those in the shaft were more than adequately insulated compared with many installations of the period.[32]

Underground they were connected to inverted T shaped bars secured to the roof. A spring-operated slide block with supported, flexible leads completed the connection to the locomotive. For ease of working a seat was placed at either end for the driver. Controls, adjacent to these seats, simply consisted of two levers; one for the brake and the other for starting or stopping. The latter incorporated the reversing facility.

Detailed tests were made to evaluate the locomotive's performance. The highest duty cycle returned was 6·4 hp when running fifteen loaded tubs at 3 metres per second. Further tests, based on these results, returned an electrical transmission efficiency of 46·4%. Subsequent calculations showed that the cost of working 660 wagons over a sixteen-hour day was just under £1.

Despite the arduous working conditions, mechanically the locomotives were successful, although teething problems were experienced with the spur-gearing and the slide block. A more robust gearing was designed and wheels were fitted to the slide block to reduce the friction.

Other Siemens & Halske installations followed at the Consolidated Paulus & Hohenzollern collieries (1883) and New Stassfurt colliery (1884). Essentially they were the same as at Zaukeroda. The sliding block was used at the Hohenzollern Colliery and again gave frictional problems. The introduction of an oil-box to provide track lubrication only exacerbated the problem, as the oil oxidised to form a very effective insulating layer!

Current pick-up was a matter of concern and ingenious methods were devised to improve this. Notable were the wide roller and the bow trolley.[33] The former involved supporting above the locomotive an aluminium roll of approximately two inches diameter, extending almost the whole width of the track. The advantages were that the locomotive and trolley-wire did not so readily part company, particularly on bends, and wear due to sparking was distributed over a much wider area. The bow trolley ultimately became more popular. This consisted of a rectangular bow of stout wire which came into contact with the underside of the wire. Unlike the trolley pole, it was not necessary to turn it around whenever the locomotive changed direction.

In the autumn of 1890 AEG installed an experimental locomotive at the Richterschacht 1 coalmine in Upper Silesia.[34] Although it was similar in appearance to that of Siemens & Halske, the method of current collection differed. Only the positive rail ran overhead; from this the current was taken to the locomotive via a solid metal block under spring pressure. The current return was through the track itself. Although accounts stress its success (running costs were 5·7 pfennigs per ton-kilometre) the installation was removed within twelve months because of 'lack of interest' of the colliery officials 'in trying to overcome such difficulties as always present themselves with new arrangements'. Probably they met opposition from the workers who feared that mechanisation would threaten their livelihood. This was an understandable reaction and became more prevalent as more mining machinery appeared over the years.

6.2.2 Running costs

In 1887 Siemens & Halske published comparative costs (correct to 0·01%!) relating to their electric locomotives working in the three collieries mentioned earlier.[35] Conveying costs varied considerably but direct comparisons are not possible on account of local conditions. A better assessment can be made by comparing the costs of different haulage methods in the same colliery. These figures are summarised in Table 6.3.

Table 6.3   Comparison running costs for various methods of conveying coal (1887)

Method of haulage	Cost of conveying (pfennig per ton-kilometre)		
	New Stassfurt	Zaukeroda	Hohenzollern
Electric locomotive	12·9314	9·1885	6·7380
Man-power	34·2	21·0	18·0
Horse-power	16	12·2	10·0

These collieries were pleased with the performance of the locomotives. By October 1891 Zaukeroda had two, Hohenzollern and New Stassfurt four. Presumably the wheeled sliding block or *kontaktwagen* had become standard.

6.2.3 Electric locomotives in British collieries

By 1890 the electric locomotive had proved itself in underground applications on the Continent, in the United States and elsewhere. The same could not be said of Great Britain. By then one electric locomotive was working in the British coalfield, at the Wharncliffe Silkstone Collieries, although H. Ravenshaw, one-time colleague of Goolden and Atkinson declared at the beginning of this century that he 'cannot quote a single instance in which electric locomotives have been employed in coal-mines.'[36] In discussion of Ravenshaw's paper, Professor C. le Neve Foster (HMI) seemed to confirm this by mentioning their use only on the Continent, in America and Vancouver.[37] Foster made passing reference to an interesting accumulator-powered locomotive in northern France, where purging was used as a means of operating in fiery environments. Purging involved pumping compressed air (or carbonic acid gas) into a motor or other item of electrical equipment. The pressurised air escaped through the gaps and flanges in the casing and so prevented the ingress of any inflammable gas.

At least one other locomotive was operating in a British mine by 1891, but the mineral extracted there was silver lead ore.[38] This locomotive, at the Greenside Mine, Patterdale, in the English Lake District, received a 250 V dc supply, collected through four contact-pulleys. It ran on a 22 inch gauge and was capable of pulling a total of 18 tons of ore 'with the greatest ease'.

This locomotive dispensed with six horses undergound, with much lower maintenance costs: less than £25 was expended in three years, including accidental breakages and brush replacements. The mine management believed this confirmed the success of the venture, particularly when they had no professional electrician on their staff. The most notable feature of the whole system, however, was that the electricity was generated by water power.

Before considering the Wharncliffe installation in more detail it is worth summarising the inherent problems with existing haulage systems. This will help to explain why British collieries did not embrace a form of mechanisation whose effectiveness, under certain conditions, was beyond doubt.

Although underground rope haulage systems were generally reliable and increased outputs, they were wasteful in manpower, especially where tubs were coupled into trains, where different rope systems converged, or where changes in direction or gradient had to be negotiated. For main haulages, locomotives were more flexible and they could be easily adapted for changing conditions by altering the car make-up and so optimise outputs.

Locomotives had some disadvantages, most directly related to the physical and geological conditions exisiting in mines. Constraints included the dimensions of the roadway gradients, which often varied considerably in length and steepness along the same length of track, curve radii, and the condition of the floor upon which the track was to be laid. Such difficult conditions were prevalent in British pits. These, coupled with the very real dangers in gassy pits of electric shocks or explosions from uninsulated cables, ensured that electric locomotives could rarely operate to maximum advantage, except on specially constructed roadways in gas-free mines.[39]

Immisch & Co. were responsible for introducing an electric locomotive into the Wharncliffe Silkstone Collieries.[40] The locomotive, named the *Precursor*, was fitted with a 4 hp motor, powered by accumulators having sufficient charge to last eight hours. Messrs. Immisch announced that during the three month trial period, 'the locomotive as a whole has given better results than was anticipated". These were guarded comments, but it was the first self-contained electric locomotive set to run under such trying conditions of space, power and weight. The main problem lay in short-circuiting of cells, but that Immisch & Co. were 'giving the matter their constant attention' and looked forward 'to a large extension of electric haulage in the near future.'

The haulage system at the Wharncliffe Silkstone Collieries was expanded in 1890[41] and again Immisch & Co. were involved, although earlier that year they had become part of the General Electric Power Company Limited. They now used a locomotive of an entirely different and new design. The rope was fixed at either end and then passed over a sprocket wheel or friction-clutch geared to a motor on the locomotive, which could then haul itself along the rope. The locomotive did not have to rely on the friction grip between its wheels and the track, so inclines and brows presented no problems. This locomotive was not designed to replace horse haulage but negotiate the short, steep lengths on the roads which horses found difficult. The 10 hp locomotive motor was connected by a pantograph to an overhead dc conductor (trolley-wire), with the track as earth return. Bonding of the rails at the joints ensured good electrical conductivity.

The trolley-wire drive on electric locomotives and vehicles in general became very popular. It was simple in design and gave high efficiency with relatively low maintenance costs, which offset the inconvenience of having to reposition the pantograph when changing direction (at least for the trailing arm type). For low roads, where it was impractical to run trolley-wires, locomotives powered by storage-batteries were used.[42] Although the idea of a completely 'self-sufficient' locomotive held considerable appeal, as it must have done initially at the Wharncliffe Collicries, the disadvantages associated with the batteries severely restricted their application. A large number of cells (each producing, typically, 2 V) were required to power a 200/240 V locomotive. Whether these were built into the locomotive or drawn behind in a separate wagon, the result was the same — a reduction in payload. The size of the locomotive motor was effectively limited to about 8 hp. Batteries needed recharging regularly (how often depended upon load and track gradient) and the maintenance costs increased after about two years.

A compromise design, developed for use in low headings or where it was not possible to run trolley-wires, cleverly combined the advantages of the trolley-wire and the storage-battery. In the main roadways it ran off the trolley-wires, while in low headings the pantograph was lowered and the locomotive switched over to storage-batteries.

Specially designed for working low headings was the Reeling or Gathering Locomotive,[43] of which the Jeffery was the most popular. The locomotive was fed from an external source by means of a flexible, double-insulated cable which was automatically coiled or uncoiled from a reel fastened behind it. The reel or drum could be either friction or power operated. This arrangement enabled locomotives to enter headings with as little as 29 inches clearance. Because of the limitations imposed by the cable and reeling, the range was restricted to 400–500 feet but these locomotives proved their worth as a shuttle between the car-loading point and the main haulage.

For use in very narrow or uneven roads, especially where sharp curves were unavoidable, the Atkinson brothers with Frederick Hurd patented in 1891 a locomotive with a flexible wheel base.[44] This maintained four wheel contact to the track at all times, and thus ensured maximum traction. A single motor was built into the locomotive frame with the armature assembled longitudinally. The wheels were driven through a set of mitre gears positioned at either end of the armature shaft. Other locomotives based on this principle made spasmodic appearances over the years. The Goodman was perhaps the most notable.

Further designs of locomotive were developed to cater for specific problems, like the sprocket type for use on roads with steep gradients.

Almost all of these designs, proved short-lived in British collieries, because of the limitations outlined earlier. Apart from a few examples of locomotives operating on the storage-battery principle, the electrically driven haulage system remained the only effective and economic means of electric underground transportation in British mines.[45]

6.3 Notes and references

1 HUGHES, L.: 'Modernising a large colliery system', *Mining Electrical Engineer*. 1924–25, Vol. V, p. 13
2 MOUNTAIN, W.C.: 'Notes on electric power applied to winding in main shafts', *Transactions of the Instutition of Mining Engineers*, 1903–04, Vol. 27, p. 142
3 HUGHES, L. *op cit.*, p. 13
4 PASQUIER, A.E. DU.: 'Electric winding engines: A comparison of systems and the influence of drum profile on the performance obtained', *Proceedings of the South Wales Institution of Engineers*, 1912, Vol. 28, pp. 199–202
5 STEVENS, R.A.: 'Ward Leonard colliery winders'. Paper presented to the 6th IEE Weekend Meeting on the History of Electrical Engineering, 1978. p. 1
6 PASQUIER, A.E. DU.: *op. cit.*, p. 181

7 MASON, E.: 'Practical coal mining for miners', Vol. 2, 1950, p. 470
8 STEVENS, R.A.: *op. cit.*, p. 2
9 BIGGE, D.S.: 'The practical transmission of power by means of electricity, and its application to mining operations, *Trans. IME*, 1891–92, Vol. 3, p. 294. A fuller description of this installation is given in chapter 3
10 *Mines Inspectors Reports*. No. 7 District (Liverpool), 1900, p. 20
11 MOUNTAIN, W.C.: *op. cit.*, pp. 163–164
12 *MIR*. No. 1 District (East Scotland), 1904, pp. 25–26
13 LUPTON. A., ASPINALL PARR, G.D., and PERKIN, H.: 'Electricity as applied to mining', 1906, p. 154
14 MOUNTAIN. W.C.: *op. cit.*, pp. 142–153
15 *ibid.*, pp. 153, 161, 162
16 *ibid.*, pp. 156–157
17 HAY, D.: 'Recent electrical development in coal mining with special reference to safety problems', *MEE*, 1923–24, Vol. 4, p. 58
18 *ICS Reference Library, Vol. 37, Winding and Haulage*, 1907, s70, p.14
19 *ibid.*
20 PASQUIER, A.E. DU.: *op. cit.*, p. 192
21 *ibid.*, p. 193
22 BRAMWELL, H.: 'Re-sinking and re-equipping the Great Western Colliery Company's Maritime Pit', *Proc SWIE*, 1908–10, Vol. 26, p. 147
23 PASQUIER, A.E. DU.: *op. cit.*, p. 196
24 BRAMWELL, H.: *op. cit.*, pp. 142–143
25 *ibid.*, pp. 150–151
26 HUGHES, L.: *op. cit.*, p. 13
27 *The Electrician*, 12 May 1911, pp. 132–135; 18 August 1911, pp. 754–756. The Old Duffryn, Abercwmboi and George Pit were quoted as owned by the Markham Steam Coal Company. These collieries were, in fact, part of the Powell Duffryn Steam Coal Company.
28 NELSON, R · 'Electricity in coal mines: a retrospect and a forecast', *Journal of the Institution of Electrical Engineers*, 1939, Vol. 84, p. 602
29 PASQUIER, A.E. DU.: *op. cit.*, p. 171
30 NELSON, R.: *op. cit.*, p. 602
31 FORSTER, B.R.: 'The subterranean electric railway at Zaukerode', *Foreign Abstracts of the Proceedings of the Institution of Civil Engineers*, 1882–83, Vol. 73, Pt. 1, pp. 474–479
32 In the shaft the central copper conductor was insulated by a layer of india-rubber within a lead pipe. This was surrounded by another layer of rubber and an outer casing of tinned iron wires. Yet more protection was provided for by a tarred linen wrapping
33 WALKER, S.F.: 'Electricity in Coal Mining', p. 347
34 EILERS, K.: 'Electric locomotives in German mines', *Transactions of the American Institution of Engineers*, 1891, Vol. 20, pp. 357–368
35 *ibid.*, p. 365
36 Remarks made by Professor C. le Neve Foster in Discussion of RAVENSHAW: 'Electricity in mines;, *JIEE*, 1900–1901, Vol. 30, p. 859
37 *ibid.*, pp. 859–860
38 BORLASE, W.H.: 'History and description of the Greenside silver lead mine, Patterdale', *Transactions of the Federated Institute of Mining Engineers*, 1893–94, Vol. 7, pp. 645–649
39 *Special Rules for the Installation and Use of Electricity issued by the Home Office under the Coal Mines Regulation Act, 1887, (Adopted 1905)*. Section V1: Electric Locomotives — specifies the conditions under which locomotives may, or may not, be used. Regulation 2 effectively restricts locomotives operating off overhead conductors to main roadways by stating 'that trolley-wires must be at least 7 feet above the level of the road or track'
40 SNELL. A.T.: 'An electric locomotive for mines', *Midland Institute of Mining Engineers*, 1887–88–89, Vol. 11, pp. 333–336
41 Albion T. Snell outlined the expansions at the Wharncliffe Collieries during the Discussion on 'Electric Mining Machinery', *ICE* 1890–91, Vol. 104. Pt. 11, p. 116. The prototype locomotive is described in *Improved System of Haulage*, Patent No. 20,070, 1889

42 *ICS*, 1907, s78, pp. 1–2
43 *ibid.*, pp. 30–31, see also MASON, E.: *'Practical coal mining for miners'*, 1950, Vol. 2, p. 515
44 Patent 13,926 1891, *'Improvements in electric locomotives and in electric conductors for use therewith'*
45 NELSON, R.: *op. cit.*, p. 603

Chapter 7
Safety, legislation and education

7.1 Introduction

Towards the end of the nineteenth century the use of electricity in mines was increasing and, in the absence of any established codes of practice for safe *and* reliable installation by either the equipment manufacturers or the coalowners, its use in potentially hazardous environments caused widespread concern. As electricity began to be used in applications other than lighting and small power, the threat took on an added dimension. The terrible death tolls from earlier explosions, although not then caused by the use of electricity, were a salutary reminder of the danger of anything that could ignite gas.

The two main hazards from electricity underground were well recognised, although satisfactory methods of overcoming them were still some years away: open-sparking and electric shock. Open-sparking (i.e. a spark capable of igniting gas) was considered by many a constant danger as the early motors were dc and had brushes and windings exposed. Collieries were reluctant to use them, even in mines generally considered gas-free.

Subsequent improvements in motor design, including the encasing of brushgear in cast-iron 'flameproof' boxes, and even totally enclosed motors failed to dispel the doubts of many colliery owners where the slightest suspicion of a gassy seam existed. Consequently motors were located in a proven 'safe area' of the workings, normally the main roadway or the bottom of the downcast shaft. If this was not possible, they preferred to use steam or compressed air for motive power, even though these were more expensive, inefficient and inconvenient to manage. Many collieries persisted in this until well after the introduction of the polyphase motor. The term 'flameproof' was not defined until 1924, when certified testing commenced at Sheffield University.[1]

The other area of concern, electric shock, centred around the problems of efficient earthing. Earthing was a contentious issue and engineers debated long and hard as to what was best. Should the installation be earthed or non-earthed? If earthed, should earth-plates, single earth or multiple earth points be used? This will be discussed more fully later in the chapter.

Ignorance remained a stumbling block in the electrification of collieries. The majority of colliery managers did not have staff with sufficient expertise to specify, install and maintain equipment. Manufacturers and consultants could not always be relied upon for pertinent advice or help because they had little knowledge of the mining industry and the conditions under which machinery was expected to work. As a result, personal opinion and preferences, rather than proven engineering principles, influenced the development of far too many early systems. Very few collieries, even within the same group, were able to boast systems fully compatible with one another.

There can be little doubt that many early installations were of inferior quality and built with scant regard to safety. The installation at the Cannock Chase collieries in 1886 certainly fell into this category. The author of a paper describing the Cannock installation was diplomatic in his choice of words when referring to their practice of utilising as electrical cables old materials, such as pit ropes, rails, water and gas mains.[2] He effectively said that the colliery management was not afraid to experiment in the name of economy. If their choice of conductors was disturbing, then their method of installation was even worse. Underground the ropes (conductors) were simply wrapped with old brattice cloth or tarpaulin; on the surface they were laid in brick chambers filled with gas tar and coaldust.

There was no legislation to cover standards of installation at that time, but in 1882 the Institution of Electrical Engineers (then the Society of Telegraph Engineers and of Electricians) published their *Rules and Regulations for the prevention of fire risk from electric lighting.*[3] Whilst neither statutory nor specifically directed at mining applications, they offered a good, commonsense approach to wiring and highlighted potential sources of danger. Fire was the major hazard, caused primarily by short-circuits and poorly made joints. Out of twenty-one rules, eleven were devoted to wiring techniques. The remaining rules covered dynamos, lamps and danger to personnel.

7.2 The 1902 Inquiry

Recognising the problems and dangers inherent in the unsystematic development and usage of electricity in mines and the growing public concern over the possible contribution of electricity to disasters at the turn of the century, the Home Office ordered a Departmental Inquiry in 1902. A six-man committee was set up, under the chairmanship of Henry H.S. Cunynghame CB, to:

> 'inquire into the use of electricity in Coal and Metalliferous Mines and the dangers attending it; and to report what measures should be adopted in the interest of safety by the establishment of Special Rules or otherwise.'

56 witnesses were called to give evidence, ranging from eminent engineers to colliery managers, colliery engineers and miners' agents. All were subjected to exhaustive and searching questions regarding their attitudes to, and experience of, electricity. Opinions varied considerably as to what constituted good and safe practices. Some, like W.C. Mountain, had definite views (some of which are discussed later) and their opinions carried much weight. Others spoke more from conviction: D. Watts Morgan, Miners' Agent in the Rhondda Valley, South Wales, was in favour of coal-cutters and mechanisation generally, provided proper precautions were taken, but conceded that his experience of electricity had not extended beyond the use of batteries. It was no easy task for the committee to sift through such diverse views and produce suitable rules. They did, however, and in the preamble to their Report, published in 1904, made two positive and all-embracing statements regarding the use of electricity in mines:

> 'It is obvious that if an agent so potent as electricity is installed with

insufficient skill, and handled with carelessness or ignorance, accidents may result, but if *properly set up* [present authors' italics] and used, it presents, we think, no features of such danger as would justify its prohibition . . .

'When therefore we reflect on the mitigation of severe and exhausting labour, and the far greater efficiency that the use of electricity is likely to produce, and which ought to result in benefit to the miner, the consumer, and the coalowner alike, we have no hesitation in saying that this new agent ought to be welcomed. The representatives of the owners and of the miners fully appreciate these views, and of the numerous witnesses who appeared before us there was not one, whether owner, engineer, manager or miner who offered the least objection to the use of electricity when *proper precautions were taken* [present authors' italics].'[4]

'Properly set up' and 'proper precautions were taken' were undoubtedly the key words, and to this effect 115 rules, divided into eleven sections, were drawn up. These embodied four general principles[5].

(i)   Electrical plant should always be treated as a source of potential danger.
(ii)   The plant, in the first instance, should be of thoroughly good quality, and so designed to ensure immunity from danger by shock or fire; and periodic tests should be made to see that this state of efficiency is being maintained.
(iii) All electrical apparatus should be under the charge of competent persons.
(iv)   All essential apparatus which may be used when there is a possibility of danger arising from the presence of gas should be so enclosed as to prevent such gas being fired by sparking of the apparatus. When any machine is working every precaution should be taken to detect the existence of danger, and on the presence of gas being noticed, such machines should be immediately stopped.

In the view of the committee the cause of most accidents could be traced to 'bad plant or arrangements'.[6] Unsatisfactory plant was often due to electricity having been introduced in instalments or on a small scale — an inadequate commitment. Included in this category were temporary installations, because, once such an installation had proved itself, there was rarely any attempt to uprate it to its newly acquired permanent status. Under-powering of equipment received much criticism, usually allied with the 'temporary' installation on a 'try it and see' basis.

Unsatisfactory plant and underpowering accounted for most problems in the use of electricity. They arose from the largely uncontrolled early growth of electricity.

On the topic of electricians and competence, interesting exchanges took place during the Evidence stage between Alfred Tallis, General Manager of the Tredegar Iron and Coal Mining Colliery Companies (TIC), and the committee.[7] The TIC was a large concern, second only to the Powell Duffryn Company in South Wales, and had five collieries producing a total of 1,100,000 tons of coal per annum. Their electrical installation was according to W.C. Mountain, whose company installed the equipment: 'what I believe to be the largest colliery installation of three phase ac work in the UK.'[8]

Tallis asserted that the men running the motors were selected from the better men originally employed on the haulage gear, which was driven by compressed air and, earlier, by steam.[9] These men received a week's training and were then given the responsibility of running the motors. They learned 'on the job' really. He was convinced that skilled electricians would not have been more suited for the job and, to support this claim, he pointed out that the plant had operated under such an arrangement for the last eighteen months.

In answer to the question, 'They are mechanicians first?', Tallis replied:

> 'With all due respect to electricians we find they are better electricians than they are workmen — they know too much.'

The interviewer evidently understood the basis of such a response and asked if Tallis wanted a mechanician first and an electrician afterwards. Tallis's reply was in the affirmative.

The TIC employed one skilled (trained) electrician, although his responsibilities covered all the collieries in the group. Tallis believed the mechanicians were 'quite capable' of doing any ordinary repair to any electrical plant in the colliery. Often they, in turn, would delegate maintenance tasks, where no specialised knowledge of electricity was required, to 'practical electricians'. Tallis asserted that these, too, were better than their technical counterparts for colliery work. The skilled electrician had a part to play in directly supervising the mechanicians if the task or problem was complicated. Such working arrangements were by no means uncommon and the committee undoubtedly took note of such practices when formulating their rules.

It is interesting to note that the term 'electrician' was not used; merely a 'competent' person. In this respect, the Rules stated[10] that:

> 'the manager shall appoint a competent person or persons to inspect daily all electrical machinery and apparatus . . .
>
> 'A competent person shall be on duty at the mine when the electrical apparatus or machinery is in use.'

This regulation added that, when the amount of electricity installed underground exceeded 200 bhp, two competent people were required; one underground and the other on the surface.

The Rules compiled by the committee were re-drafted and reduced from 115 to 56 and enforced in 1905.[11] They were the first legislation designed to promote the safe use of electricity in mines. They allowed sufficient latitude for the competent engineer to develop his skills and expertise. Unfortunately, what constituted a competent engineer was not defined and, with a dearth of such people, poor installation and standards were likely to continue.

7.3 A time for consolidation

Commendable though the efforts of the committee were in formulating the 1905 Rules and correctly predicting the rapid growth of electricity, they perhaps failed to appreciate just how rapid that growth was to be. One recent commentator has criticised the weighting of the six-man committee since only one was an electrical engineer.[12] If more had been invited to participate, perhaps the report would have had a different emphasis. That is pure conjecture, but the state of contemporary knowledge on the use of electricity underground contained many 'grey' and uncertain areas. As these areas included the construction and installation of equipment underground, cable construction, earthing techniques and the dangers of gas or coal dust ignition, the resultant rules needed using with extreme caution.

It was unfortunate that the increasing use of electricity coincided with an escalation of explosions and accidents throughout the coalfield. Although there was no proof whatsoever that electricity was to blame, public opinion began to turn. Gaps in electrical knowledge and the fact that the Royal Commission, set up in 1906 to investigate coal dust explosions, after two years still could not agree upon the mechanics certainly did not help. At one stage things became so emotive that the representatives of the Miners' Federation in the House of Commons demanded that the use of electricity underground be prohibited.[13]

It is easy to paint a picture of gloom and pessimism. Problems did exist, but in reality the situation was never that bad. Many colliery managers continued to choose electricity as the most logical means of modernising their collieries. Where the larger concerns, such as the Powell Duffryn, took such a step, they usually achieved safe *and* reliable installations. Of 7,248 fatal accidents underground between 1905 and 1910, seventy were caused by electricity. These figures go a long way towards putting the situation into a proper perspective.[14]

The problem lay with the smaller collieries which had neither the capital to invest in first-class installations nor the expertise to select or maintain the system. Such concerns tended to 'cut corners' and opt for dc drives, poorly designed systems and unarmoured cable.[15] They often relied upon advice from their own 'engineers', who, more often than not, were mechanics who had 'grown-up' with the system. To such men the Rules meant precious little.

In 1908 the *Electrical Review* carried a scathing article following a fatal accident at one such colliery.[16] It drew attention to the:

'serious inefficiency of persons in charge of electrical installations at collieries, and to the disgraceful and negligent way in which many of these plants are installed and attended to.'

The fatality occurred at Dark Lane Colliery, Yorkshire, when an electrician came into contact with a live 'pull wire', which controlled a sluice valve on a coal washing machine.

In the ensuing inquiry HM Chief Inspector for the district was asked by a juror why he had not inspected the electrical apparatus at the colliery. He replied that 'it was impossible for him to do so' because he and a fellow inspector had 400 mines and 800 quarries in their district alone. Regrettably this situation prevailed throughout the rest of the country: there were too many mines and not enough inspectors.

Since the introduction of the Regulations in 1905, from 3,000 mines there had been 1,158 notifications to District Inspectors regarding the installation of electricity.[17] Recognising the sheer volume of work and the limitations of inspectors who were not electrical engineers, R.A.S. Redmayne (HM Chief Inspector of Mines) received permission from Mr. Gladstone, the Prime Minister, to appoint an *electrical* inspector of mines, because it:

'became desirable therefore that I should have at hand at the Home Office a qualified electrical engineer who would keep me informed of developments, and who would make regular tours of inspection of the electrical plant at the mines and report to me thereon.'[18]

Following consultations with C.H. Merz, Robert Nelson was appointed the first electrical inspector of mines in November 1908.

In spite of Nelson's ability and enthusiasm, single-handed he could not hope to monitor and effect the changes so obviously required. The best he could do was to conduct another survey and, in the light of its findings, revise the 1905 Rules. Nothing less would do than a thorough revision of these rules, which had been formulated on the technology and experience of lighting and small power drives. Now three-phase ac systems formed the core of many installations, driving heavy winders, conveyors and coal cutters. There had been a massive development in the distribution and use of electricity underground. Nelson's revision came in the form of the 1909 Inquiry.

7.4 The 1909 Inquiry

In the Autumn of 1909 a three-man committee was appointed to: 'consider the working of the existing special rules for the use of electricity in mines'. This time two electrical engineers were nominated, C.H. Merz and Robert Nelson. The third person was R.A.S Redmayne himself.

The subsequent Inquiry was thorough. Over fifteen days the committee called 36 witnesses and asked 6,250 questions. The questions were comprehensive and covered all aspects of the use of electricity in mines. Again, people involved in mining, manufacture and education were called. A number of witnesses at the earlier 1902 Inquiry were recalled, and their responses made possible accurate assessment of the growth and problems particular to specific collieries or organisations.

Just after Christmas the draft Report was submitted to an eminent barrister, who had a knowledge of electricity, to make it watertight, as Redmayne said, in interpretation without sacrificing clarity.[19] It concluded[20]:

'Apart from the more concise arrangement and wording of the Revised Rules, it will be seen that the main directions in which the requirements are to be met have been strengthened:

(i)   By prohibiting the use of electricity when, on account of the risk of explosion, such use would be dangerous.

(ii)   By providing that inflammable material shall not be used in the construction of motor rooms where there exists the risk of fire.

(iii) By more stringent regulations as regards the earthing of the outer coverings of apparatus.

(iv)   By clearly setting forth the conditions to be fulfilled by switchgear.

(v)   By insisting upon the better mechanical construction of cables and apparatus.

(vi)   By providing for the proper supervision of apparatus.'

As in the earlier Inquiry the committee adopted a favourable attitude to the use of electricity in mines. Its use was to be encouraged, provided proper consideration was given to the selection, installation, maintenance and supervision of electrical equipment.

There was little doubt in the committee's mind that it was in the construction and maintenance of electrical equipment that the biggest *and* most effective improvements could be realised. They emphasised this:

'. . . the committee is convinced that a high standard of construction is an economy, not an expense, proper supervision, . . . is still necessary. With well constructed apparatus the necessity for the periodic examination of working parts remains, and the absence of repairs which should ensue must be credited jointly to both good construction and proper supervision.[21]

The earlier rules had not actually prohibited the use of electricity in any part of the mine, irrespective of working conditions. Now the committee advised amendment: electricity should be excluded from some parts of the mine. The determining criteria were to be whether, during daily inspections, in any one place quantities of inflammable gas were detectable with an ordinary safety lamp. Where gas was seldom found, 'provided good and proper maintenance are assured so as to obviate as far as is practicable the risk of open-sparking',[22] no objections could be raised to its use. The level of inflammable gas at which electricity was to be shut off was specified at $1\frac{1}{4}\%$.[23]

Since the formulation of these Revised Rules coincided with the passage of the Coal Mines Act (CMA) through Parliament, the former was incorporated into the latter under 'General Regulations' and came into force in 1913.[24]

A very important point was raised during the evidence stage of the Inquiry, namely, that the Home Office ought to take the responsibility of approving apparatus to be used in mines. The committee declined to accept this suggestion because[25]:

(i)   Official approval of apparatus would hamper progress.

(ii) Approval of each individual piece of apparatus would be involved for the system to work efficiently, for as much depends upon the way in which the parts of electrical apparatus are fixed together as upon their design.

(iii) Such approval might engender less consideration on the part of the management of a mine in introducing electricity, and less care in the maintenance and operation afterwards.

The reasoning can be understood, but hindsight suggests that an excellent opportunity was lost. A 'central approving authority' might have been established at a time when 'new' regulations and safety concerns were a topic of attention. The Americans had already established their Bureau of Mines, to act as an official approving body. In Germany, in 1904, the Westphalian Mining Association was promoting the use of electricity in mines and testing existing designs to determine their suitability in inflammable atmospheres.[26] There can be little doubt that such concerted efforts reaped dividends and helped those countries to become more technically advanced than Britain. Some of the major British manufacturers took it upon themselves to do testing, but were handicapped for lack of comprehensive facilities. No certificates were issued with the equipment.

For the first time the term 'electrician' appeared in legislation, 28 years after appearing in the first edition of the IEE Regulations. Regulation 131(*b*) stated:

'An electrician shall be appointed in writing by the Manager . . . .'[27]

and Regulation 118 duly defined an electrician as:

'a person appointed in writing by the Manager to supervise the apparatus *in the mine* and the working thereof, such person being a person who is over 21 years of age, and is competent for the purposes of the rule in which the term is used.'[28]

(Note: these regulation numbers relate to the General Regulations of the Coal Mines Act).

The word 'competence', together with the possibility of issuing statutory certificates of competency, was discussed in some detail during the Evidence stage. The committee decided against insisting on certificates on the grounds that:

'. . . however well certified a man may be, considerations of habit and character are of even greater importance in this connection than technical knowledge.'[29]

Not underestimating the importance of safety, they continued:

'Safety and efficiency would be more likely to be promoted by the employment of a conscientious and resourceful man than by a man who may have the advantage as regards technical knowledge but who is inferior as regards thoroughness.'

The onus of deciding whether a person was competent was left to the colliery manager.[30] A manager with a certificate of competency would have attained a reasonable level of electrical proficiency, but might not have kept up with the continual developments in the electrical field, so the task of deciding upon an electrician's competency was not easy.

7.5 Cables and earthing

During the Evidence stage of the 1902 Inquiry it became abundantly clear that ideas as to what constituted the ideal cable differed considerably. W.C. Mountain, of the Ernest Scott & Mountain Co., Newcastle-upon-Tyne, represented the interests of the Mining Association of Great Britain at the Inquiry. He considered vulcanised cable was the most suitable for hostile environments.[31] Vulcanised bitumen and lead covered cables, he maintained, could be affected by water. For three-phase installations, he preferred to run single-cored cables, but a single, three-cored cable was acceptable, if suitably armoured.

Other engineers advocated vulcanised rubber, or lead covered cables, or any permutation. A few were bold enough to suggest that bare wires supported on insulators, provided that they were in full view, were more than adequate.

Cable damage had always been a major problem and one of the main safety concerns. Earlier, in 1902, a series of experiments were undertaken at Llwynypia Colliery, Glamorgan, to ascertain the durability of armoured cable when installed underground. The intention was to submit the findings in evidence at the 1902 Inquiry. To simulate roof falls, stones of various shapes and sizes were dropped directly onto cables laid on the ground. Tests took into account varying floor conditions and also involved fixing cables in different positions in timbered headings. The extent of the damage, in all cases, was evaluated. Briefly, the findings were:[32]

- That a greater likelihood of cable damage existed from falls of comparatively small, sharp stones than much heavier smoother ones.
- Damage to cables was significantly reduced when in timbered headings.
- Also, in timbered headings, where mechanically hauled trams were worked, cables should be placed near the ground. This method was preferred to running the cables close to the roof because of the dangers of 'pile-ups' from runaway trams.
- Wet mines aggravated cable inquiry.

To further complicate the issue, opinions varied on the most suitable working voltage level.[33] A safe maximum was considered to be 500–600 V, because of the perpetually damp atmosphere and coal dust accumulation on machinery. Coal dust accumulations were a problem on cables because, in the event of a short-circuit, the vibrations disturbed the coal dust and a potentially explosive environment was created around the faulty portion of the cable.

The wide diversity of opinion on cables alone illustrates well the situation which had developed through letting 'personal preference' reign unchecked. It was no easy task for the Inquiry to formulate regulations that would satisfy everyone.

However, they ruled that bare conductors were only acceptable on surface applications and provided that they observed regulations relating to minimum height and methods of fastening. Unless lead armoured, all had to have further protection, and in damp situations this was to take the form of metal tubes, which were to be watertight, electrically continuous and efficiently connected to earth.

Many of the observations made in the Llwynypia experiments formed the basis of regulations for installing cable underground. Where high voltage (greater than 650 V) or extra high voltage (greater than 3,000 V) applications were intended, the cable had to be tested at twice the working pressure. The tests had to be backed by a written certificate, signed by the cable manufacturer or competent engineer of the *United Kingdom*[34] [authors' italics].

J.H. Fisher, of the Lambton Collieries, patented about 1908 a device which ensured that, unless the earth connection to a piece of equipment (a coal cutter in this case) was intact, then the supply could not be switched on.[35] However, a special trailing lead was required, having five cores (earth, pilot and phase connections). The monitoring circuit, which consisted of a transformer and coil, was mounted in the gate end box. The five core cable connected between this switch and the coal cutter to complete the circuit. If the earth connection became defective for any reason, the coil de-energised, tripped the switch and isolated the machine. This simple system proved successful and was the forerunner of more modern and sophisticated sensing circuits.

About this time, individually screened cables were first advocated, but they were not readily available until after the First World War.

Many engineers saw the cable as the most vulnerable part of the system (a view shared by the 1909 Inquiry). Despite constant efforts to achieve adequate earthing, fatalities still occurred. An incident in 1909, involving a three-phase, three core cable at the Cribbwr Fawr Colliery, Glamorgan, graphically illustrates the engineers' dilemma.[36] A 550 V bitumen insulated, armoured cable, which supplied pumps and a small haulage set underground, was damaged when a number of trams 'ran wild'. The electrician 'cut back' the armouring, repaired the cable and earthed the outer armouring. Unknown to him, the cable had sustained other damage further along its length. A collier received a fatal shock when he used the cable as a handrail to alight from a carriage. Soon afterwards another collier came into contact with the cable: he too died.

The subsequent inquest revealed that the system of earthing at the colliery was in sections, and the dryness of the ground rendered this form of earthing ineffective. To safeguard against similar tragedies in the future, the company modified their earthing system by laying large copper sheets in a brook near the surface of the colliery. As a further safeguard, the cable armouring was made electrically conductive across all junction boxes.[37]

Local earth plates, used underground, were considered unsatisfactory by the 1909 Inquiry, as they had been known to fail[38] through dryness, as above, or even through chemical corrosion caused by the sulphur content in the coke,[39] while mechanical damage was always possible. Achieving a good earth was not always easy, as Robert Nelson admitted, but he had never found a case where this was insuperable.[40] He did not favour a single earth plate, since an efficient earth could not always be guaranteed. Where such doubts existed, the sinking of a second earth plate, within twenty yards or so of the first and connected to it, normally solved the problem. Nelson stated that such a configuration, in good earth, was nearly as effective as a single earth plate in the same surroundings with a surface area equal to the two.

The Powell Duffryn Company ensured a high degree of safety with respect to earthing, installing local earth plates to back up the main earth.[41] The 1909 Inquiry rightly recommended that the existing rules should be made more stringent as regards to earthing.[42] They believed that the risks of fire and explosion due to small leakage currents could only be avoided by ensuring a proper connection to earth for the metalwork of all equipment and armoured coverings on cables which carried current at a potential greater than low pressure (i.e. 250 V). Without such a connection, an open-sparking or dangerous condition could arise in the time that it took for the earth leakage protection to operate. Because of the widely varying working conditions experienced in mines, with their wet and dry areas, the safety margin between a safe or unsafe condition was so little that earthing was essential to avoid the risk of electric shock.

The Inquiry did not insist upon earthing of the neutral point on polyphase systems. This met with mixed reactions. Sparks, in his 1915 IEE paper, was adamant that this should be made compulsory at the next revision,[43] to ensure the impossibility of continued system operation when a definite fault existed. Other engineers, however, liked to have an unearthed neutral point so that system operation would remain unaffected until a second such fault developed.

7.6 Electrical fatalities

In a paper read before a joint meeting of the Scottish branches of the National

Association of Colliery Managers and the Association of Mining Electrical Engineers in April 1911, Robert Nelson stated that more than 70 fatalities had occurred underground over the period January 1905 to December 1910 as a result of electricity.[44] Electric shock was by far the largest category — 55 deaths in 53 accidents. There were 13 deaths from the ignition of firedamp and two from underground fires.

Although Nelson produced a detailed breakdown of the 53 accidents, a summary of them (Table 7.1) clearly demonstrates the trend.[45]

Table 7.1   Summary of fatalities from electric shock (Jan. 1905 to Dec. 1910)

	Accidents	Deaths
Absence of, or inefficient earthing	22	24
Defective insulation of cable system	20	20
Contact with uninsulated live parts	9	9
Miscellaneous causes	2	2

Nelson was uncompromising in his attitude to these statistics and must have caused some uneasiness amongst managers (or at least those present) when he asserted that two fifths of these deaths could have been avoided if there had been an efficient connection to earth. The full implication of his words could scarcely have gone unheeded:

'that such a provision [i.e. inefficient earthing] coupled with proper attention to the insulation of those cable systems upon which accidents occurred, would have avoided not less than *four-fifths* of the total number of accidents.'

As for those who came into contact with uninsulated live parts, Nelson thought this figure would diminish as the design and manufacture of equipment improved.

Nelson's conclusion was predictable: 96% of these accidents were avoidable. It was a sobering thought.

In the same paper Nelson produced evidence, conclusive to his mind, to settle the long standing debate over armoured or unarmoured cable. He obtained his 'evidence' from Scotland where this cable controversy attracted a great deal of attention.[46] His findings are summarised in Table 7.2:

Table 7.2   Accidents relating to the use of armoured and unarmoured cables in Scotland (Jan. 1905 to Dec. 1910)

	Accidents	Deaths
Armoured cables	2	3
Concentric cables	2	2
Unarmoured cables	20	20

The first two kinds could have been avoided had an efficient connection to earth existed. Not everyone accepted the case for armoured cable and, in the discussion which followed Nelson's paper, Alexander Anderson disputed Nelson's claim on the grounds that the use of unarmoured cable underground exceeded that of armoured in the ratio 12½ (and not 10 times as Nelson had stated).[47] That aside, the fatality ratio remained the same. Anderson disagreed on a number of other points:[48] that he did not believe that concentric cable could be regarded as a protected cable, and he thought that Nelson should advocate lower pressures underground, e.g. say, 100 V. Nelson asked Anderson under what circumstances would he neither earth the system nor use unarmoured cable.[49] Anderson immediately countered that he did not disbelieve in earthing, but had based his comments on the experiences of a seaside colliery where, because of the effects of saline, the cable had to be replaced within six months. Anderson believed that under these circumstances an unarmoured cable coated with a suitable preservation compound would have been more suitable. Sensing the growing disagreement between Nelson and Anderson, Robert McLaren, HM Inspector of Mines, entered this discussion, suggesting that perhaps a test case would soon arise and effectively settle the matter.[50] Nelson, having the last word, thought that Anderson had neglected one very important consideration — that of providing protection from mechanical damage.

In relating fatalities to voltage levels, Nelson made some interesting observations. Forty-five fatalities had occurred on medium pressure systems (650 V and below), four on low pressure systems (250 V and below), and four on high pressure systems (650 V and above).[51] Again the discussion revealed a diversity of opinion. A.B. Muirhead thought that, if the manufacturers and users of electricity were compelled to classify everything above 250 V as high tension, the medium as well as high pressure fatalities would be a thing of the past.[52] Muirhead's line of reasoning was sound, but scarcely that of a practising engineer because more stringent restrictions regarding isolation etc. would make cumbersome the majority of straightforward, daily maintenance tasks. Muirhead added, perhaps unfortunately, that Nelson was concerning himself too much with earthing and too little with insulation.

7.7 Senghenydd: an avoidable tragedy?

Two firedamp explosions occured in 1912, both caused by sparks emitted from electric bell signalling systems. These were the first in a sequence of events which cumulated in the Senghenydd disaster of 1913.

It has always been hotly disputed whether the Senghenydd disaster could have been avoided, if the coal owners had heeded a circular issued by the Mines Inspector following an explosion at Bedwas Colliery, Monmouthshire, the previous year. One recent writer had little doubt, describing the disaster as 'caused by the culpable negligence of the owners and their agent.'[53]

The firedamp explosion at Bedwas Colliery injured twelve miners, three of whom subsequently died from their burns.[54] One of the injured later recalled how, before losing consciousness, he looked across to the bell when it rang and saw a flash of flame. The inadvertent shorting together of the wires by men working elsewhere in the colliery caused the bell to operate.

Experiments performed later verified that the current produced by an emf of 11·5 V (the value measured across the Leclanché cells after the explosion) was sufficient to ignite an explosive mixture of lighting gas and air. Even when the potential was reduced as low as 4 V, ignition of the mixture was still possible.[55]

Recognising the dangers from open-sparking on signalling systems in gassy conditions, W.N. Atkinson, Mines Inspector in charge of part of the South Wales Coalfield, circulated the following letter to all South Wales colliery owners, detailing the Bedwas incident and urging strict observance of Electrical Special Rule 15(1):[56]

123 Cathedral Road
Cardiff

28th August 1912

Gentlemen

I am directed by the Secretary of State to inform you that an explosion of firedamp occurred recently in the South Wales Inspectors Division, by which 12 men were burnt, 3 of them so severely that they subsequently died.

The explosion was proved beyond all reasonable doubt to have been caused by the sparking of an electrical signalling bell which ignited an accumulation of gas resulting from a derangement of the ventilation due to the breakage of air pipes.

It was afterwards proved experimentally that sparks from the bell in question worked by a battery of 11·5 volts would ignite an explosive mixture of lighting gas and air, and the mixture was also ignited by sparks from signalling wires produced by a current of only 4 volts pressure.

I am instructed therefore to call your attention to the necessity of strict observance of Electrical Special Rule 15(1) with reference to signalling apparatus.

Yours faithfully,
W.N. ATKINSON

Electrical Special Rule 15(1) read:

> 'All cables, apparatus, signalling wires and signalling instruments shall be constructed, installed, protected, worked and maintained, so that in the normal working thereof there shall be no risk of open-sparking.'[57]

The second explosion occurred at Caebontpren Colliery, Carmarthenshire, after bell wires were rejoined by mechanics who had just completed the testing of the system. This time the consequences were less severe: four men received burns.[58]

Atkinson, in his report for that year, briefly recorded details of both accidents and commented that he believed the explosion at Bedwas was: 'the first recorded case of a colliery explosion caused by sparks from signalling apparatus'.[59]

He also pointed out that electrically operated signalling systems were being installed much nearer to the coalface than in the past — with the consequent heightened risk of firedamp explosion. Once again he drew attention to the need to observe Electrical Special Rule 15(1), especially in fiery mines.

Senghenydd stood at the head of the Aber valley, Glamorgan, just a few miles north of Bedwas. It was a new colliery — the first coals were raised in 1895 — but already it had gained the reputation of being one of the more dangerous mines in

Britain on account of its frequent outbursts of gas. Many people were convinced that it was only a matter of time before another, perhaps major, explosion occurred. They did not have to wait long. On Tuesday, 14 October 1913, 439 lives were lost.

Although insufficient evidence was available at the inquest to determine the cause of the explosion, the Coroner returned a verdict of accidental death on the miners.[60]

Open-sparking was discussed as a possible cause of the explosion, but the jury thought that the weight of evidence suggested ignition by a naked light in the lamp station.

The jury added their recommendations, one of which was:

> 'We are of the opinion that there are not sufficient Inspectors of Mines to enable a thorough inspection of the collieries to be made, and as often as the work ought to be done.'

A statement that had a familiar ring!

7.7.1 A disquieting laxity

At the end of January 1914, the formal inquiry recommenced and in April its findings were made public. There had been breaches of the practice required by the Coal Mines Act (1911) and some of the findings were at variance with those of the inquest. Moreover, Redmayne's conclusions differed from those of his fellow assessors.

The most serious breaches concerned the inability to reverse the air flow through the workings immediately and the control of coal dust. Other breaches were less serious, but when considered as a whole, they pointed, in Redmayne's words, to a 'disquieting laxity in the management of the mine'.[61]

As for the actual explosion, Redmayne believed that a large volume of gas was released following a fall of rock in the Mafeking Incline.[62] The striking together of falling rocks or a spark from the electric signalling system used on the haulage were the only apparent means of ignition. Explosions were known to have originated from both these sources in the past. The only other possible means of ignition was from safety lamps or matches, but no sign of either was found near the place of the explosion, and vigorous searches of the collier (at the pit head) further reduced the possibility of matches being taken underground.

After quoting the General Regulations relating to precautions against open-sparking from electrical equipment in mines having inflammable gas, Redmayne regretted that a better method of excluding sparks had not been adopted. Astonishment was expressed that the management should have taken the risk of sparks igniting the gas after the Bedwas Colliery explosion: 'which proved beyond reasonable doubt to have been caused by the sparks from an electric bell.' He drew attention once again to Dr. Atkinson's circular letter.

The electric signalling system installed at the Mafeking Hard Heading (the probable cause of the explosion) was one of several in the colliery[63] used to let the engineman know the state of readiness of a haulage. A cast iron, non-gastight casing housed the trembler-type bell located in the enginehouse. Power was applied to the circuit by Dania batteries. To make a signal, a miner would physically short together with a metallic object the twin bare wires, fixed 12–18 inches apart on the side timbers. This produced sparks at the make-and-break contact on the bell.

Although the bell was housed in a non-gastight case and bare cables were used, this did not violate the new Special Rule 15(1) or any other regulation because, ironically, signalling systems installed prior to 1 June 1911[64] were exempted. Consequently, this one was covered by the regulations of 1905. The relevant regulation said:

> 'In any place or part of the mine where General Rule No. 8 of the Coal Mines Regulation Act, 1887, applies, bare wires shall not be used for signalling circuits *except* [authors' italics] in haulage roads, and the pressure shall not exceed 15 V in any one circuit.'

The continued use of such installations was permitted until 1 June 1920 unless an objection was raised by the Inspector of the District. There had been no objection.

Redmayne recommended that the regulations be amended to ensure that *all* signalling systems be treated like any other piece of electrical equipment where gas may occur — as a potential source of danger.

7.7.2 Further investigations

Voltage readings taken by the men responsible for the maintenance of the signalling system indicated 9 V — this was confirmed by H.W. Scheilbach, consulting engineer of the Lewis Merthyr Consolidated Collieries Ltd.[65] Dania batteries after use had the characteristic of being able to recover to virtually their former potential — a fact confirmed by the battery manufacturers. On this basis it was surmised that after an extended period, overnight, for example, the battery potential could well have exceeded 9 V. Further doubt was cast, when it was found that the readings were taken on a linesman's detector. Although the competence of the men who took the readings was not questioned, tests carried out by the Board of Trade (BoT) showed that some expertise was required to read the instrument accurately, particularly when the readings fell between gradations. Under these circumstances, it was possible for an error of ±1 V to occur over an aggregate of nine readings.

In order to clarify what was now becoming a crucial issue, tests were carried out by Robert Nelson at the New Tredegar Rescue Station, Monmouthshire.[66] Using a sample of gas from the Cymmer Colliery he was unable to obtain ignition by breakflash (the spark produced when a short was removed from the wires) below a potential of 11·9 V. This result seemed to cast doubt on open-sparking as the cause of the explosion, even though it clearly contradicted the results obtained after the 1912 Bedwas Colliery explosion.

Redmayne, however, did not consider these tests sufficiently exhaustive and asked Dr. R.V. Wheeler to undertake further investigations at the Home Office's Experimental Station at Eskmeals, near Barrow-in-Furness. Wheeler conducted his tests with equipment of the same specification as that used at Senghenydd and produced a spark capable of igniting a methane–air mixture with a methane concentration of 8·2%. The current was 450 mA and the voltage 4·5 V (measured on open-circuit). Open-sparking also occurred at the bell trembler mechanism, although here the current was 700 mA and the voltage 7·5 V. Wheeler concluded:[67]

(i) That methane–air mixtures vary in their ease of ignition according to the percentage of methane present. The lower and higher limits being 5·6% and 14·8% respectively, with the most sensitive band laying between approximately 8% and 9%.

(ii) The inductance of the circuit was of prime importance in determing whether the spark produced at breakflash can cause ignition.

(iii) The current was of greater importance than the voltage.

Redmayne described Wheeler's findings as 'very important'[68] and the possibility that the explosion was caused by open-sparking probably remained in his mind.

7.7.3 A final confirmation?

Even after the Inquiry was completed and the contents fully digested, the matter did not rest. Testing of signalling systems continued at the Eskmeals Station under the guidance of Wheeler. The object was to obtain a bare wire system that eliminated the danger of breakflash ignition of firedamp. They were successful and the ensuing report, published in January 1915, accurately detailed the limiting parameters of methane levels, self-inductance, current and voltage.[69] Recommendations were made for modifications to bell systems to ensure safe working even under the most sensitive conditions.[70] Most importantly, the report confirmed the findings on breakflash conditions that were earlier presented to the Senghenydd Inquiry.[71]

After this no doubt could have remained in anybody's mind about the dangers of electricity in mines. Lessons were learned, but at high cost in human suffering.

7.8 Training and education

7.8.1 Association of Mining Electrical Engineers

The scathing article, mentioned earlier (p. 139), which appeared in the *Electrical Review* in 1908, also carried some constructive comments about the need to form a responsible society. This society, as the *Electrical Review* explained, could appeal to all mechanical and electrical engineers and, like the Colliery Managers Association, provide mutual benefit and protection. The article went as far as proposing a name for this society — The Association of Colliery Engineers. To ensure that such an important and pressing suggestion did not become yesterday's news, the *Electrical Review* invited correspondence from their readers.

The response was encouraging; there was no doubt that the *Electrical Review's* thinking was correct. In March 1909 a letter from J.C. Williams of Whitehaven, entitled 'Institution of Colliery and Mining Electrical Engineers', appeared in their columns.[72] Williams saw the over-riding aim of the proposed institution as being to: 'decrease the number of electrical fatalities in collieries.'

Williams suggested that such an aim could be realised through: 'lectures, discussions and practical suggestions to keep up a high standard of workmanship and materials best suited for colliery work . . .'

Membership would comprise, principally, working colliery and mines' electrical engineers or electricians — a wider range than that suggested by the *Electrical Review*.

Williams put a lot of thought into his letter and produced a set of rules around which the institution could develop. Safety, arising from good working practices, clearly dominated his thinking. All aspiring members were to demonstrate their competence in a number of prescribed ways; they were to furnish references, demonstrate previous experience (either in charge for one year or an assistant for two years of a colliery electrical plant), pass an examination set by the institution, and satisfy the committee absolutely that he: 'is thoroughly experienced and fully competent to take charge of any colliery or Mining plant.'

His concern regarding competence can be understood. Many people held similar views on the importance of 'certified competency', but the Committee in the 1909 Inquiry thought otherwise, believing that strength of character outweighed certification.

Williams's letter provided the necessary impetus. In April 1909 a meeting was held at the Grand Hotel, Manchester, to discuss the formation of an institution.[73] Over fifty engineers and other eminent people attended, and the Institution of Mining Electrical Engineers was founded. William Maurice was chosen as first president. A better choice could hardly have been made, because Maurice, after qualifying as an electrical engineer, had been a colliery electrical engineer and then a colliery manager.

Almost immediately the Institution took positive steps to enforce competency, since this was central to their philosophy. They were particularly concerned with the competence of those *actively* engaged in the care or management of mining electrical installations.[74] To this end they set their own examinations with Professor Thornton of Armstrong College, Newcastle, as Chief Examiner and Robert Nelson on the examining board.

William Maurice, in his Presidential address (see chapter 1), outlined the aims of the Institution. These aims differed considerably from the practically orientated ones suggested by Williams. Maurice was seeking to establish a new class of professional mining electrical engineers.

Safety remained essential, and the Institution should 'promote the adoption of approved methods and devices tending to increase safety.[75]

Right from the outset the Institution demonstrated its willingness to become involved and made an important contribution when William Maurice gave evidence to the 1909 Inquiry.[76] He had circulated the paper, which he intended submitting to the Inquiry to members and his evidence took account of their views.

The need for such an Institution and an indication of its future success can be gauged from the membership which by 1913 had grown to 1,083. It had changed the name to Association but had no doubts about the way forward, as W.C. Mountain indicated in his Presidential address in 1913 (see chapter 1), even though it meant a great deal of hard work and endeavour on everybody's part.

There is little doubt that the Association, together with other learned societies and institutions, brought about an increase in technical knowledge, awareness and competence. The existence of these bodies, however, could not, and did not, guarantee an end to all problems. Problems still existed, particularly in the smaller collieries, and continued for a very long time. Their 'engineers' were not attracted towards learned bodies or societies.

7.8.2 Mining schools

Up until the time that Redmayne accepted the Chair of Mining at the newly established University of Birmingham in March 1902, only two premier institutions in this country possessed schools of mining, the Royal School of Mines at South Kensington and the College of Physical Science at Newcastle.[77] This contrasted sharply with North America where a large number of mining schools were established. The most prominent were:[78]

In the United States
Colorado State School of Mines
Columbia University
Harvard University (Lawrence School of Mining)
Lehigh University (specialising in coal-mining)
Massachusetts Institute of Technology
Michigan College of Mines
Ohio State University (specialising in coal-mining)
University of California (college of mining)
University of Minnesota
University of Missouri

In Canada
Kingston School of Mines
McGill University (Montreal)
Toronto University

Joseph Chamberlain, the Chancellor of Birmingham University, displayed considerable acumen by suggesting that his Professors of Mining and Metallurgy visit these American and Canadian centres of learning to study their organisation and facilities, and recommend any features that might be incorporated into the University's proposed new buildings at Edgbaston.[79]

No time was lost in making arrangements, and in early June Professors Redmayne and Turner set sail for North America. What immediately impressed Redmayne was the practical approach adopted by the Americans. They endeavoured to create as far as possible, whether in the laboratory, classroom or field, the conditions under which the student would be expected to work. Redmayne discussed these techniques in depth with Professor Bovey of McGill University and several other professors. The consensus of opinion was:

> 'that the majority of the failures in mining work in Canada and the Western States of America were attributable to English mining engineers refusing to adopt American methods.[80]

Redmayne apparently remained non-committal on this, but conceded that Great Britain 'did not lay great stress on collegiate training for practical men.' North American mine owners would no longer consider employing a person unless he held a diploma from a leading institution.

Redmayne managed to visit several copper mines, including Calumet and Heela in northern Michigan, and he was pleasantly surprised to find that most of the mine captains were Cornishmen. He took the opportunity to talk to these men who, he hastened to add, had adopted American mining techniques. Whatever was said certainly impressed him because he later wrote:

> 'They [the mine captains] did not fail to impress me upon the necessity of adopting American mining practice in the British Empire, if we would hold our own in world competition.'[81]

By the end of this fact-finding tour there was no doubt left in Redmayne's mind that technical education and mining practices in North America were well in advance of those in Britain. He summed everything up effectively:

'This superiority, I came to realise, was in large measure due to the fact that they sought to exemplify and enforce precept by practice in their educational scheme . . . .[82]

Along such guidelines he prepared his syllabus for the School of Mines at Birmingham University.

The course of study at Birmingham University lasted three years. Two were devoted to the principles of pure science and the final year to applying these principles. Predictably, practical work played a significant part and Redmayne's persuasions must have impressed the Chancellor of the University who, in the degree ceremony of 1903, was prompted to say:

'And we propose to institute what, I believe is a new experiment in this connection . . . an underground model of a mine, where all the problems connected with mining, with the work of a mine, with underground surveying, and with ventilation, can be explained and studied.'[83]

Not an entirely new idea, but one which was certainly very unusual and one which caused a great deal of interest both in Britain and abroad. A model mine had been constructed in the United States since Redmayne's visit; on the Continent one was especially constructed for the Great Exhibition held in Paris three years earlier; and in the southern hemisphere there was one at the Ballarat School of Mining in New South Wales, Australia.[83] The only parallel in Britain was in Cornwall where the Camborne School of Mining, in 1897, had taken over the eastern part of the South Condurrow Mine in Cornwall.[84]

The model coal-mine at Birmingham which was constructed out of brick, covered three-quarters of an acre and 'sunk' ten feet below ground level was opened in 1908.[85]

Redmayne's syllabus proved so successful that, with minor modifications, it was adopted by many of the leading technical institutions — a very apt accolade to a man who had a major influence in setting and maintaining standards of excellence in the British coalfield.

7.9 Notes and references

1 BAKER, F.H.: 'Background to the Coal Mines (Mechanics and Electricians) General Regulations, 1954', *The Mining Electrical and Mechanical Engineer*, May 1957, p. 305
2 The talk given to the British Association by A. Sopwith was reported in *The Engineer*, 1886, Vol. 62, p. 402 and *Engineering*, 1886, Vol. 42, p. 310
3 BOWERS, B.: *A history of electric light and power*, Peter Peregrinus, Stevenage, 1982. The regulations are reproduced in full on pp. 215–218
4 *Report of the Departmental Committee on the Use of Electricity in Mines* (hereafter referred to as *1902 Report*), London, 1904, p. 7
5 *ibid.*, pp. 9–10
6 *ibid.*, p. 8

7 *Minutes of Evidence taken before Departmental Committee on the Use of Electricity in Mines* (hereafter referred to as *1902 Evidence*) London, 1904, QQ1763–1964, pp. 59–65
8 *ibid.*, Q4, p. 2
9 *ibid.*, Q1861, p. 62
10 *1902 Report:* Rules for the Installation and Use of Electricity in Mines, Section 1 (General), Rules 20 and 21, p. 24
11 BAKER., F.H.: *op. cit.*, p. 304
12 LOYNES, E.: *'The road to progress with safety'*. The story of the Institution of Mining Electrical and Mining Mechanical Engineers, Manchester 1984. Here the author states on p. 10 'It is noteworthy that only one member of this Committee was an electrical engineer.' He emphasises the point again on p. 145: 'The regret was that only one of the six members of this committee was an electrical engineer'
13 *ibid.*, pp. 11–12
14 NELSON, R.: 'Avoidance of electrical accidents in mines'. Paper read to a joint meeting of the Scottish Branches of the National Association of Colliery Managers and the Association of Mining Electrical Engineers, *Mining Electrical Engineer*, 12 April 1911, pp. 371–377
15 *Report and Evidence relating to the Working of the Special Rules for the Use of Electricity in Mines* (hereafter referred to as the *1909 Inquiry*) London, 1911, Q6145, p. 210
16 This article is reproduced in LOYNES, E: *op. cit.*, pp. 203–205 and TOMOS, D.: *'The South Wales Story of the Association of Mining Electrical and Mechanical Engineers,'* Cardiff, 1960, p. 10, the latter being in its original form
17 REDMAYNE, Sir R.A.S.: *'Men, Mines and Memories.'* 1942. p. 164
18 *ibid.*
19 *ibid.*, p. 165
20 *1909 Inquiry*, pp. 11–12
21 *ibid.*, p. 12
22 *ibid.*, p. 7
23 *'The Law relating to mines under the Coal Mines Act, 1911'*, London, 1914, Section 60(2), p. 71.
J. Samuels in his Presidential address to the South Wales Branch entitled 'The electrification of mines, and the demand for technical training'. *Mining Electrical Engineer*, 1921–22, Vol. 2, p. 98 states that following discussion the level of inflammable gas at which electricity was to be cut-off was reduced to 0·5%. However, the *CMA Sectionalised*, 7th edition (revised 1937), p. 121 still quotes the cut-off level as 1·25%
24 BAKER, F.H.: *op. cit.*, p. 305
25 *1909 Inquiry*, p. 7
26 SAMUELS, J.: *op. cit.*, p. 101
27 BAKER, F.H.: *op. cit.*, p. 305
28 *ibid.*
29 *ibid.*
30 *ibid.*
31 *1902 Evidence*, pp. 3 and 6
32 *ibid.*, Appendix 13, p. 224
33 *1902 Evidence*, QQ8–9, p. 2
34 *1902 Report*. Rules for the Installation . . . Section 111 (Cables), Rule 4, p. 25
35 LUXMORE, S.: 'Early applications of electricity to coal mining', *Proceedings of the Institution of Electrical Engineers*, 1979, Vol. 126, p. 872
36 LEWIS: *Mines Inspectors' Reports*, 1909, pp. 68 and 85
37 *ibid.*, p. 68
38 *1909 Inquiry*, p. 10
39 NELSON, R.: *op. cit.*, 'Experience related during the Discussion by Mr. H.A. McGuffie', p. 380
40 *ibid.*, pp. 376–377
41 SPARKS, C.P.: 'Electricity applied to mining'. *JIEE*, 1915, pp. 398–402. He discusses earthing principles in considerable detail
42 *1909 Inquiry*. p. 9
43 SPARKS, C.P.: *op. cit.*, p. 402
44 NELSON, R.: *op. cit.*, pp. 371–373
45 *ibid.*, p. 374

46 *ibid.*, pp. 374–375
47 *ibid.*, p. 380
48 *ibid.*
49 *ibid.*, p. 383
50 *ibid.*, p. 382
51 *ibid.*, p. 375
52 *ibid.*, p. 381
53 NOEL, G.: *'The Great Lockout of 1926'*, 1976, p. 19
54 ATKINSON, W.N.: *MIR*, 1912, pp. 7–8
55 *ibid.*, p. 8
56 *Report of the Inquiry into the Senghenydd Explosion* (hereafter referred to as the *Senghenydd Inquiry*), 1915, Appendix E, p. 54
57 *Senghenydd Inquiry*, p. 31. This rule became General Regulation 132 (1) when incorporated into the CMA
58 ATKINSON, W.N.: *MIR*, 1912, p. 8
59 *ibid.*
60 BROWN, J.H.: *'The Valley of the Shadow*, Port Talbot, 1981, p. 122
61 *Senghenydd Inquiry*, p. 35
62 *ibid.*, p. 31
63 *ibid.*, p. 6
64 *ibid.*, p. 35
65 *ibid.*, p. 19
66 *ibid.*, Appendix B, Table B, p. 60
67 *ibid.*, p. 36
68 *ibid.*
69 WHEELER, R.V.: *'Report on battery-bell signalling systems as regards the danger of ignition of firedamp–air mixtures by the break-flash at the signal-wires'*, 1915, p. 6
70 *ibid.*, p. 11
71 *ibid.*, p. 2
72 LOYNES, E.: *op. cit.*, pp. 208–209
73 *ibid.*, p. 17
74 *ibid.*, pp. 26–27
75 MAURICE, W.: 'Presidential address, AMEE, Vol. 1, 1909–10, p. 15; see chapter 1 for fuller extracts from this speech
76 LOYNES, E.: *op. cit.*, p. 147
77 REDMAYNE, R.A.S.: *op. cit.*, p. 43
78 *ibid.*
79 *ibid.*
80 *ibid.*, p. 44
81 *ibid.*, p. 46
82 *ibid.*, p. 47
83 *ibid.*, p. 49
84 *ibid.*
85 *ibid.*, p. 50

Chapter 8
Private generation

8.1 The way forward?

The necessity for greater economy in coal production caused the industry to look at better and more effective methods. Some of the more progressive colliery owners believed that one way to achieve these objectives was through a modernisation scheme involving the use of electricity.

Immediately this raised the question of whether to purchase electricity from a power company or to generate one's own? There was, of course, no stock answer and each colliery had to be considered on its own merits. Charles P. Sparks, a prominent electrical engineer, said in a paper to the IEE in March 1915:

> 'Speaking generally, when a supply is available from a power company it should pay individual collieries to purchase the electrical energy required'.[1]

One of the reasons he gave was to:

> ' . . . avoid adding to the already multifarious duties of the colliery management the necessity of supervising an electrical power station.'

Such a statement was not without foundation. At that time, very few collieries had staff with sufficient expertise to carry out relatively simple maintenance, let alone run a complex generating plant.

Sparks was responsible for successfully master-minding the largest colliery electrification scheme in the country for the Powell Duffryn Steam Coal Company (PDSC) (later dealt with as a case study). He continued with a self-evident statement:

> 'Where, however, the requirements of a single colliery undertaking exceed the output of the smaller power companies, it is advisable for such undertaking to generate its own electricity supply.'[2]

It might, indeed, be a group of collieries, as in the case of the PDSC.

At the same meeting C.H. Merz endorsed Sparks's comments on supplies purchased from an electric power company. Looking, no doubt, at the wider possibilities with characteristic vision, he added:

> 'in every case where a connection to an electric power company's system is possible it should prove advantageous, even if only as a standy-by.'[3]

Having established this pertinent point, Merz asserted that at very little extra cost the transmission capacity in this country, with a few exceptions, could be doubled, or even trebled. He thought that, where private and public lines covered the same ground, it must be profitable to join them, provided, of course, that the systems were compatible.

8.2 A very difficult decision

However the colliery owners arrived at the most economical solution, the answer had to be worked out in terms of cost per ton of coal produced. This figure had to include all costs, not just the direct ones. Often the indirect or 'hidden' costs were, at best, educated guesses, and, at worst, pure hunches.

Many factors were involved in formulating the cost of electrification, but the starting point had to be a consideration of the local conditions. Did the colliery have fairly new and efficient steam winders? Was the colliery an individual undertaking or part of a larger group? If the latter, were there other collieries geographically near? Did the lines of either a power company or municipal supply run nearby?

Many collieries would hesitate to go 'all electric', but might compromise and utilise exhaust steam from the winders to drive mixed pressure turbo-alternators. A good case could be made for installing such equipment when the quantity of exhaust steam was just sufficient to provide electricity to supply all the auxiliary loads (an average of 45 lb of exhaust steam per horsepower hour could produce 1 kWH of electrical energy). Additional exhaust steam could be obtained from ventilation fans and compressors.

Not everyone, however, liked this approach. One person who had reservations was C.H. Merz.[4] In the Discussion following Sparks's paper, Merz declared that electrical engineers could 'confidently look forward' to the time when advances in technology would justify the complete electrification of *all* [present authors' italics] colliery plant. The exhaust steam turbine was just a temporary expedient. Or, as he put it: 'a useful means of getting over a transition period, but nothing more.'

Sparks, replying to Merz's comments,[5] agreed that future simplification of electric winding plant would reduce initial costs even further, and in larger collieries steam would eventually succumb to electricity. However, he still saw a widespread use for exhaust steam turbines, particularly in smaller collieries geographically isolated from the lines of a power supply company. This was not an unfair comment considering the conditions prevailing in South Wales, an area with which Sparks was more than familiar.

Winders, especially the ac variety, presented the biggest single problem towards complete electrification because of the peak loads at the commencement of the wind. For an individual colliery wanting to erect its own generating station, cost would be inordinately high unless loads such as fans, pumps and compressors could significantly improve the load factor. Ideally such loads would account for 50–60% of the turbine output, with the peaks of the two winders superimposed on this. Winders fitted with equalisation or operating on the Ward-Leonard principle had a lower peak current but these had other drawbacks. Such considerations were often sufficient to deter individual collieries from generating their own electricity and make them look instead at bulk purchase from a power company or municipal supply, always provided that transmission lines ran nearby and electricity could be purchased for less than a penny per unit.

8.3 Fuel costs

Many of the larger collieries, groups and combines, like PDSC and the Tredegar Iron Company (TIC), both in South Wales, had little hesitation in generating their own electricity and, whilst not all were able to match the cost of an established *and* efficient power company, they chose to do so in order to remain totally self-sufficient. This attitude laid a new emphasis on the word 'power' and did not make the progress towards a nation-wide supply of electricity any easier.

As regards costs, large collieries possessed a number of inherent advantages, the most obvious being fuel. They could reduce this item to a trifle for the coal was literally available for picking up at the pit mouth. They used little more than small and other unsaleable coals. If the generating station was within the confines of the colliery complex and sited near the washery, transportation and handling costs were also minimal. Coal conveyors improved the situation even further. In the case of a central generating station, aerial ropeways kept this cost to a minimum.

Pressing home the advantage of cheap fuel, collieries not only used exhaust steam but also gases from blast furnace and coke ovens, to the chagrin of one power company, the South Wales Electric Power Company (SWEP), who moaned about such 'unfair advantages' to anybody willing to listen. A case of sour grapes perhaps, but if the large collieries had turned to SWEP for power it was extremely doubtful whether they could have provided a service of the reliability demanded. SWEP's system was poorly engineered, with the lines strung out over a wide area with small and widely scattered plant. Their 25 Hz supply frequency conflicted with the 50 Hz and 40 Hz systems that the PDSC and TIC respectively used. As regards poor engineering, SWEP was certainly not alone; the Cleveland and Durham, and the Derby and Nottingham schemes experienced similar problems.[6]

Despite protestations, the fuel costs of power companies were not inordinately high because of the increased thermal efficiency of their boilers. This was achieved by burning high grade fuel and reducing steam consumption of the larger generators by using higher pressure steam.

A common practice was to allow for a high rate of depreciation of equipment in order to write off capital as quickly as possible. By taking these factors into account, in real terms, the difference in costs between a private generator and a power company was not too great.

8.4 Electricity and legislation

Before considering the relationship between private generation and the electric power companies, it is necessary to appreciate how the supply network developed within the legislative structure.

In the years around 1880, various local authorities submitted Bills to Parliament seeking authority to supply electricity, and the Government deemed that a general Act was necessary to provide guidelines for future applicants. Consequently the 1882 Electric Lighting Act was passed. Under its terms the Board of Trade (BoT) was empowered to grant a Licence or Provisional Order to any local authority, company or even individuals, authorising them to supply electricity within a specified area. Licences were granted to interested parties for a period of seven years (renewable), provided that the consent of the local authority was forthcoming. Almost without exception permission was refused, which resulted in the BoT having to issue a Provisional Order (PO), which required Parliamentary confirmation.[7] The disadvantage of a PO was that the local authority had the right, after a period of twenty-one years, to purchase the undertaking compulsorily. The restrictions imposed by this clause have often been blamed for the reluctance of companies to invest in generating electricity. Although the Government of the day was criticised by Emile Garcke and others for imposing such severe restrictions on private enterprise, later commentators have argued that twenty-one years was more than enough to see a return on investment.[8] This may well be the case, but some sympathy must go to those municipalities supplying large rural areas. Lack of technical knowledge and expertise often restricted the growth of an effective distribution system.

Although different sources give different numbers of Provisional Orders and Licences granted in the years immediately before and after 1882, they all clearly demonstrate the lack of response displayed by investors. Probably the most reliable and most quoted figures were those included in a reply that the President of the Board of Trade (who administered the Act) gave to a Common's question in February 1888. He stated that, although some 64 franchises (59 POs and five Licences) had been granted to companies and some 17 franchises (15 POs and two Licences) to local authorities, none had exercised their rights.[9]

Towards the end of 1884 further pressure was put on the Government to amend the 'purchase clause' of the 1882 Act and after protracted discussions the Second Electric Lighting Act came onto the statute books in 1888.[10] This was a short Act, being really an amendment to the Act of 1882. The period of tenure was doubled — from twenty-one to forty-two years — unless the Provisional Order specified a shorter period. There was still no stipulation regarding 'goodwill', should the local authority wish to enforce compulsory purchase. A public supply authority had no reason to be complacent since the Act provided that the granting of a Provisional Order was not exclusive: competition was always a real threat.

The new Act had the desired effect and increased confidence all round. This, in conjunction with technical advances, enabled electricity supply schemes to grow rapidly. By January 1889 there were twenty-six central generating stations in operation and a further seventeen under construction. Within twelve months this had risen to 46 and by January 1891 the figure reached 54.[11]

8.5 Power companies

By about 1900 technical developments had assured the commercial viability of high voltage ac transmission over long distances. This, together with the availability of larger and cheaper generating plant, encouraged the formation of power companies. Power companies, incorporated through Special Acts, were protected against purchase from the local authorities and had rights extending over large areas, but some restrictions were imposed on their operation. They could provide electricity in bulk to authorised distributors and power direct to other consumers, but problems arose if the latter lay within the boundaries of a local authority who had powers of supply established before the formation of the power company. Under such circumstances the power company could only provide a power supply with the consent of the local authority. Furthermore, they could only supply electricity for lighting if the consumer agreed to take a power supply.

The attitude of local authorities towards the formation of power companies was mixed. Some, such as Durham and the North Metropolitan, who provided power to authorised distributors only, had no problems. Those, however, in the coal mining areas of South Wales and Lancashire experienced considerable opposition from the large municipalities who believed that they alone should have the right to supply power in their areas. Provided the large municipalities could show that they were able to provide power as cheaply as the company undertakings, they were allowed to supply power as well.[12] Thus the undesirable situation arose where areas supplied by the power companies were broken up by the municipal supplies. It was not uncommon to see transmission lines of varying voltage and frequency running in close proximity or parallel to each other for considerable distances. In areas where private generation abounded, the pattern was further complicated, especially in South Wales. There, as late as the mid-1920s, the output from private generation was estimated to be four times more than that from public supplies.[13] Since all parties had different codes of practice and standards, this created problems when integration of the nation's supply began.

Of the fifteen power companies in existence by 1912, only the Newcastle Electric Supply Company (NESCo) was financially successful. Many factors contributed to this lamentable state of affairs: poor engineering, inadequate funding, lack of vision and the insistence of private industry to generate its own electricity. The latter was particularly prevalent in the coal mining areas of South Wales, Lancashire and Yorkshire.[14]

The level of private generation in the coalfields was far from uniform: South Wales had the highest anywhere in Great Britain. Although the South Wales Electric Power Company's area of distribution was large it served many scattered pockets of industry, thanks to their exclusion from the heavily populated areas by the municipalities. The SWEP's situation was not helped by the large and powerful companies, such as the PDSC and TIC, whose ambitious, private schemes effectively excluded SWEP from the most lucrative and developed tracts within the coalfield. This led to very strained relationships. All attempts by the SWEP to convince these concerns of the advantages and benefits of purchasing electricity in bulk fell upon stony ground. SWEP failed to command confidence and by 1909 a number of collieries had reverted to generating their own electricity, as the following extract from the annual report of the Mines Inspector for Monmouthshire indicated:

> 'Practically all the newer concerns have erected generating plants, while most of the older ones, which formerly took power from the South Wales Electric Power Company have also done so.'[15]

This situation may have been precipitated when in 1906 it was learned that the SWEP was unable to raise more capital and had come to some unspecified 'special arrangements' with their consumers in order to stay solvent.[16] Unlike some power companies they stayed in business.

8.6 NESCo and the North-East

At the other extreme, NESCo developed between 1900 and 1914 from a local lighting undertaking on the north bank of the Tyne to a highly integrated network supplying an area of more than 1,400 square miles. It provided the cheapest public supply electricity in Europe, at an average of 0.5d. per unit.[17]

By 1908 NESCo had developed into an extensive network supplying large areas of Northumberland and Durham.[18] Three power companies, NESCo, Durham Electric Power Company (DEPCo) and the Cleveland & Durham Electric Power Company (C&DEPCo), supplied electricity to the network, but NESCo effectively controlled the system.[19] NESCo had achieved this through a number of earlier amalgamations, with DEPCo in 1905 and C&DEPCo in 1906, after these companies had run into financial difficulties. In the case of DEPCo, NESCo acquired the rights to generate and distribute electricity in the county of Durham. The agreement with C&DEPCo included linking in a supply from NESCo's Carville power station.[20] Such moves increased the installed capacity of the system in 1908 to 101,950 hp, approximately one-ninth of the total power supplies in the United Kingdom.[21] Table 8.1 lists the stations jointly responsible for this output, with their operating voltages.

Table 8.1 Summary of NESCo's generating stations (1908)

Power station	Type	HP of plant installed	Voltage
Carville	coal fired	56,000	6,000
Philadelphia	coal fired	13,000	6,000
Neptune Bank	coal fired	6,800	6,000
Grangetown	coal fired	8,000	12,000
Hebburn	coal fired	4,500	6,000
Weardale	waste heat	6,650	3,000
Newport	waste heat	4,000	3,000
Blaydon	waste heat	3,000	6,000
Capacity of plant installed		101,950	

NESCo's main transmission and distribution voltage levels varied considerably: the main trunk voltage was 20,000 V, transmitted through a combination of underground cables and overhead lines. When this voltage was introduced in 1906, it was by far the highest transmission voltage in the country. The development of suitably insulated cables was undertaken by the German firm AEG, since British manufacturers were reluctant to tackle this.[22] In the Tyne and Durham and the Tees areas, the high tension distribution was at 6,000 V and 12,000 V, respectively, using overhead lines and underground cables. Power supplies were at 3,000 V and 400 V. In all cases the system frequency was 40 Hz.

Although expansion was rapid, there were some delays, as Merz himself admitted in a paper to the Iron and Steel Institute at Middlesbrough in 1908.[23] The main difficulties encountered were the alterations needed after the amalgamations because of different frequencies and voltages, and the necessity of shutting down small, uneconomical power stations. Merz wisely opted for system uniformity from the outset, although this meant heavy expenditure on frequency changing equipment to convert the 50 Hz Durham Collieries Co. supply and the mixed 25 Hz and 50 Hz supplies of the C&DEPCo to NESCo's 40 HZ supply.[24]

The consumers in the North East included shipyards, engineering works, rolling mills and coal mines, and these provided the diverse and varied loads essential for good load factors. The supply of electricity to collieries lagged behind that of other consumers, because NESCo had to obtain the necessary running powers. By 1908, however, the situation had changed: collieries with an output of 8 million tons of coal per year were either taking, or arranging to take, almost all of their power requirements from the power companies.[25] Some had winders of 1,600 hp. Merz confidently predicted that the supply of electricity to collieries would eventually exceed that taken by ship building and heavy engineering works combined. Even so, a number of collieries in Durham and Northumberland still generated their own electricity: Merz estimated that these burnt 2·5 million tons of coal per annum. He believed that:

> 'the same power could be provided electrically in a large central power station at one quarter the above consumption.'

He was probably correct, but as the engineer to the NESCo system he was prejudiced and must have seen it as 'lost revenue'.

Further expansion to the system was planned, with the construction of three power stations at Dunston, Bankfoot and Teesbridge. The first station was coal fired and the other two waste heat fired. Outputs were to be 30,000, 3,300 and 1,300 hp, respectively. The concept of utilising waste heat and waste gases from coke ovens and blast furnaces was unusual, but not new. The PDSC had successfully used waste gases from their regenerative coke ovens to power gas engines as early as 1896. In 1906, in evidence given before the Royal Commission on Coal Supplies, it was estimated that the Cleveland furnaces alone, even after supplying the stoves and blowing engines, would produce sufficient surplus gas to generate 61,000 hp of electricity continuously.[26]

Merz was a keen advocate of waste heat stations and made an impressive case for its widespread implementation in the North East. Despite the practical difficulties in positioning such stations near to sources of waste heat to minimise heat and pressure losses, he believed that such a system could be extended nationally, to obtain cheap electricity from otherwise wasted sources of energy. In the North East a colliery company and a power company first co-operated to build a station on the waste heat principle in 1905. The arrangement was between the owners of the Priestman Collieries and the Blaydon power station.[27]

In 1913 NESCo sold 232·4 million units and of this, 76·8 million units (almost one-third) was purchased by collieries.[28]

In 1912, NESCo had a generating capacity of 91 MW, which represented 42% of the generating capacity of the country's fifteen power companies. The Clyde Valley and the SWEP, located in areas of heavy industry, came next, with outputs of 27 MW and 18 MW respectively. Yorkshire was sixth with an output of 10 MW. The bottom nine power companies in the 'league table' could only manage a joint total of 45 MW. The combined output of the ten largest municipal undertakings was 332 MW: they all exceeded 10 MW, with Manchester, the largest, having an output of 59 MW.[29]

8.7   Load factors and load curves

Load factor — the ratio between the actual and maxmimum output — was a matter of utmost concern for both private and public generation schemes. The generating plant had to adequately cope with the maximum demand, but the more powerful the plant the greater would be the initial cost. It was in everybody's interest to keep capital costs as low as possible and therefore, to ensure that the actual output and the maximum demand were as close as possible.

Private generators — and collieries in particular — were able to achieve better load factors than power companies or municipal suppliers. The PDSC, for example, in 1908, returned a load factor of 37% whereas in 1910 the nearby Cardiff Undertaking could only reach 18·82%.[30] To achieve such high figures, collieries embarked upon a planned programme which included the grouping of collieries, centralised control, prudent location of interconnected, highly efficient power stations, an equally efficient transmission system and electrification of all colliery functions. These functions needed to be co-ordinated to obtain as smooth and uniform a curve as possible. This could be achieved in various ways. Coal could be raised on two shifts only, with the third shift confined to maintenance. Winding times could be planned over a group of collieries in order to 'even out' the peak demands. Water could be allowed to accumulate in the sump by day and pumping be done at night. The PDSC went one step further and, right under the nose of the SWEP, sold electricity on a non-statutory basis to the villages nestling around their collieries. This ensured that the already strained relationship between the two companies did not get any better!

A typical load curve for a large colliery system in midwinter shows an '8 hour dip' at either the afternoon or night maintenance shift, depending upon the colliery's working arrangements. The ideal solution would be to wind on all three shifts and 'flatten out' the curve, but one shift needed to be dedicated to the repair and maintenance of machinery, equipment and workings.

The load curve illustrates why municipal and power companies eagerly sought to attract collieries, particularly those with an 'idle' afternoon shift. The peak loads of a municipal supply in midwinter occurred between 3 and 7 pm when lights came on in commercial premises and factories, followed by street lights, as people made their way home on trams and began to switch on domestic equipment. The last effect was small because in the majority of houses electricity was still very much a luxury. Superimposing mining demands on such load curves would result in a much improved load factor. If the municipal undertaking or company was lucky enough to secure the load of something like a tin-plate works, whose continuous process maintains a high load factor of 70% or more, it could, in theory, obtain a load factor as high as 80%. In areas where private generation abounded, this could never happen, and the municipal and power companies were to remain the 'poor relations' for many years.

8.8   The Powell Duffryn Steam Coal Company: a 'model' system

'The most complete, certainly the largest, system of colliery electrification in this country,'[31] was how C.H. Merz summed up the PDSC electrical installation in 1915. Merz gave two reasons for this assertion: first, that the PDSC had engaged at the outset the services of a very experienced and respected electrical engineer, and, secondly, they managed to expand without once having to resort to the use of Parlimentary powers. Obviously there was more to the PDSC expansion than this, but they could never have expanded so rapidly or to such an extent had the inter-connection of their collieries depended upon obtaining Parliamentary powers. They had considerable powers of persuasion and influence.

The electrification of the PDSC collieries in South Wales was not something that just happened, but part of an on-going company policy to use the most up-to-date and cost-effective methods to maintain their position as a prime coal producer and a major influence in the South Wales coalfield. This they achieved through strong, aggressive management, backed by a very effective director network.

8.8.1 Formation of the PDSC

Life started in 1864 for the PDSC when, with a capital of £500,000, they acquired the steam coal collieries in the Aberdare and Rhymney valleys from Thomas Powell.[32] The output was then more than 400,000 tons per annum — 3·4% of the total South Wales output, including Monmouthshire. Within three years the output had doubled, mainly due to the acquisition of four new collieries.

The next few decades were difficult times for the company. Apart from trade recessions, the main threat came from the increased competition in the nearby Rhondda valleys. Here the quest was on to exploit the deep lying, rich steam coal seams and satisfy an ever increasing market in the iron and steel industry and the steam ship packets.

At first the effect upon the PDSC was minimal as coalowners wrestled with the formidable problems of deep mining. The PDSC's initial response was simply to increase coal production by employing more miners. However, a point was reached when the shallower seams in the Aberdare valley were being worked to full capacity. To remain competitive, the PDSC had either to open new collieries or improve efficiency. They did both. Their method of increasing efficiency was to introduce mechanisation, and this was reflected in their 1875 output when one million tons of coal was raised,[33] representing 7% of the total South Wales output. Powered haulages, driven by compressed air, were introduced in the main roadways of their larger collieries. Although a response to competition, this was not an impulsive move, but was based upon tests carried out in previous years to compare the relative merits of horse and mechanical power.[34] The horses were not released to pastures green: they were still required to haul the heavily laden trams from the coalface to the stations.

For the remainder of the century the PDSC continued to modernise, within the constraints of trade recessions. The main development was to install steam and compressed air drives, often from central points. Finance was provided out of profits or by new share issues.

8.8.2 A master stroke

In the opening years of the present century the PDSC, having doubtless observed keenly the increasing use of electricity in the mining industry, decided that it was time to embark upon their own extensive electrification scheme. Although they had some experience of electrification in their Rhymney valley workings the decision was not an easy one for these hard-line directors, and it must have

Fig. 8.1 Powell Duffryn Steam Co., Ltd., Middle Duffryn washery and power house, built 1905, to provide electricity to their Aberdare valley collieries.

caused many board-room headaches. There were no accepted codes of practice for choosing of suitable equipment or how to install it. Personal preference and cost were the only guides. Safety, too, left much to be desired. Should electricity be used underground at all, particularly in the fiery steam coal seams? If this was not enough, there was a distinct lack of electrical expertise amongst the colliery staff. Manufacturers and consultants could not always be relied upon as they were often ignorant of mining matters. A wrong decision could lose a considerable amount of output and capital. Realising this, the PDSC took the prudent step of engaging a very experienced and respected electrical engineer, with the appropriate name of Sparks![35] This appointment was a master stroke because within a decade under his able and enthusiastic direction the PDSC had become a 'single' system, attracting the attention of prominent engineers and scientists from home and abroad. Many went away to implement Sparks's ideas in their own existing or proposed systems.

The PDSC had extensive workings in both the Aberdare and Rhymney valleys. Although less than nine miles apart, geographical restraints made communication between them difficult. Initial electrification concentrated on developing each valley separately. It was not until 1915 that the two systems were linked by an overhead transmission line.[36]

8.8.3 Electrification of the Aberdare valley

The year 1905 saw the inception of the Aberdare valley electrification scheme with the building of a power station adjacent to the new washery at the Middle Duffryn colliery.[37] The site was deliberately and shrewdly chosen to take full advantage of the large quantities of small, virtually unsaleable coal inevitably produced. By a combination of railway wagons and overhead ropeways, which were electrically operated, they made the washery the central collecting point for these small coals. The railway sidings, in addition to their original function of moving coal, could also handle heavy machinery. Abundant supplies of water for steaming and condensing were readily available from the River Cynon.

They installed three generators, one of 2,000 kVA and two of 1,000 kVA rating, manufactured by Allgemeine Elekrizitäts Gesellschaft (AEG). The engines, built by Yates and Thom of Manchester, were steamed by ten Babcock & Willcox boilers with superheaters. Green's economisers were also fitted to raise the temperature of the feed water to 180°F before it entered the boilers.

Although these boilers were installed with hand-firing, plans were already in hand to fit an automatic firing system. This involved tipping the small coal into a hopper at the washery and conveying it by bucket-conveyor to the boilers. The conveyor also removed the ash.

This power station provided a three-phase, 3,000 V supply (3,300 V at the generator terminals) to all the PDSC workings in the valley by a combination of overhead lines and underground cables. When this network was completed, the PDSC was able to commence a programme of substituting motor drives for steam power on a large number of haulage gears, fans, pumps and other equipment. All motors rated above 50 hp were supplied directly at the transmission potential. Installations at this level of voltage, although common on the Continent, were a rarity in Britain: 500 V was the more accepted potential. The only other British exponents of high voltage were Messrs. Bolckow, Vaughan & Co. in Middlesbrough, and one or two other firms in the North East.[38] The 50 Hz frequency, adopted from the outset of Sparks's involvement (1903), gave sufficient flexibility for colliery purposes, especially as regards speed. This resulted in lower initial cost and economy of working. It gave designers of large generators more latitude by making available four speeds between 750 rpm and 3,000 rpm, compared with two speeds and a maximum of 1,500 rpm for 25 Hz generators.[39]

The total rating of the equipment was 6,000 hp and it returned a load factor of approximately 37%. About 15,000 units were generated daily at a total cost of 0·36d. per unit (running and capital costs each accounting for 50%).[40] These figures indicate an efficient and economic installation for that period. Many public and supply authorities could only stand in awe at such a load factor.

All seven collieries in the Aberdare valley received power from the central generating station at Middle Duffryn.[41] With the exception of a parallel feed linking Fforchaman, Cwmneol and Aberaman collieries to the generating station, the system has a radial design. Each colliery had its own substation which received the power and distributed it to the various drives. Surface equipment was supplied by bare copper conductors.

Only three collieries had underground drives, Fforchaman, Aberaman and Lower Duffryn. In these, twin high voltage feeds (3,000 V) were taken by special cables down the shaft to underground substations. Thence electricity was supplied directly to pumps and haulages or transformed to 500 V for small drives. Coal cutters and underground lighting were conspicious by their absence: the PDSC had still not resolved in their own minds the wisdom of using electricity near the coal face.

With the exception of two collieries which had electric winders, Abercwmboi

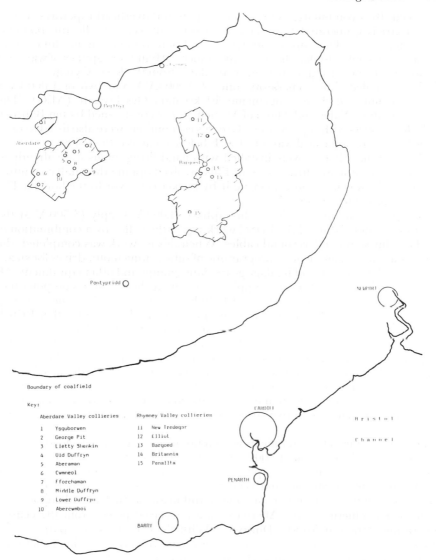

Fig. 8.2 Plan of the Powell Duffryn Collieries (1914)

and Old Duffryn and two with electrically operated fans, Old Duffryn and Lower Duffryn, the type and application of equipment in all the collieries was essentially the same, so a brief description of Old Duffryn plant will illustrate the PDSC approach. This particular colliery, however, had some unusual features not found elsewhere in the Aberdare valley complex.

An engineer who visited Old Duffryn for the first time in November 1905 recorded his initial impressions (when seeing the colliery 'on stop').

'there is an entire absence of the usual clouds of steam visible at all
times at nine out of ten of the pits in South Wales;'[42]

The reason there was no visible steam was simple; the boilers that used to provide steam for the main winders and fans had been removed. This was the first pit in the group to depend entirely upon electricity for its day-to-day running.

The winder engine, driven by a three-phase, 3,000 V, 100 hp motor was capable of raising 400 tons of coal in eight hours from a depth of 200 yards, although at the time of this engineer's visit it was only used for raising and lowering men.

The ventilation fan was belt-driven by a three-phase, 3,000 V, 180 hp motor. Its capacity was 200,000 cubic feet/min against a three inch water gauge. Since the motor was required to run continuously for long periods, a standby motor was available.

The haulage gears, of the main and tail type, were powered by 160 hp and 80 hp motors, respectively, located next to the winding house, whence ropes run down the shaft to the haulage trains underground. The PDSC, with an eye for economy, did not have new gears installed but adapted the earlier steam driven ones. Similar adaptions were made to the smaller 50 hp haulage gears which controlled the movement of the trucks to and from the screens on the surface.

Such was the extent of the Aberdare valley development that an engineer who visited the complex as part of the South Wales Institute of Engineers' tour in November 1905 pronounced:

'there can be no doubt it is by far the boldest, most comprehensive,
and interesting private installation in existence in this country or on
the Continent today'.[43]

These were flattering and laudatory words addressed to the host company, but they could not have been far from the truth.

The company was not satisfied, however, and further major work was soon undertaken. Before the electrification scheme as a whole can be assessed, the Rhymney valley developments need to be considered.

8.8.4 Electrification in the Rhymney valley

One of the first undertakings was the installation in 1904 of a three-phase, 1,000 hp generator at the Elliot colliery. This was driven by a horizontal, cross-compound engine, and both supplemented existing electrification at the colliery and provided power, via overhead transmission lines, to Bargoed colliery a few miles further down the valley.[44]

The system worked so well that within two years larger generation equipment was installed at the Elliot to ensure sufficient power for the Bargoed colliery complex, while a generating station was built there to satisfy the future needs of all their Rhymney valley collieries. The siting of this station, like that in the Aberdare valley, was deliberate: not only was it a central location, but also close to the regenerative coke ovens and a recent by-products plant, whose exhaust gases were used to fuel the prime movers.[45] This was another fine example of the PDSC's approach to utilising available resources.

The novel and successful introduction of these gas engine generators at Bargoed enabled rapid expansion to take place in the Rhymney valley.

Ultimately, the scheme provided sufficient electrical power for the complete electrification of the shallower pits and sufficient compressed air to coal cutters and conveyors in other seams. The compressed air also powered large haulages, replacing horses which, despite the 1870s trials in the Aberdare valley, still worked gallantly underground.

Additional generation was soon installed at the Elliot Pit and Penallta collieries. The latter had two 3,000 kW mixed pressure turbine sets of the Westinghouse Rateau type and one 750 kW cross compound, drop valve generator.[46] These sets fed the overhead transmission lines, which had been modified to interconnect all the Rhymney valley collieries in parallel, and ensured that each colliery received uninterrupted supply in the event of a breakdown in a generating station or feeder section. It also meant that circuit breakers, etc. could be taken out for routine maintenance.

As far as possible the PDSC opted for overhead transmission lines because, in the words of C.P. Sparks:

'Experience has proved the reliability of overhead transmission with bare conductors for colliery supply. Apart from saving in the first cost, the main risk through the use of cables is avoided, namely, subsidence.[47]

Although aluminium conductors were available and offered even further reductions in first costs, Sparks was against their use. He had used them at several collieries elsewhere but, for the PDSC scheme where conductor cross-sectional area approached 1½ square inches, he considered the problem of increased windage unacceptable. The transmission lines were protected by the Merz–Price system.

A major step in the Rhymney valley electrification programme was taken in 1910 when the PDSC sank new pits, establishing the Britannia Colliery at Pengam, a few miles further down the valley from Bargoed. These pits were sunk entirely by electric power,[48] supplied from Bargoed and Penallta. (This suggests that the generation equipment at the Elliot Pit was now only for standby.)

It has been claimed that this was the first British colliery to be sunk and worked entirely by the use of privately generated electricity.[49] This claim appears to be justifiable. It was a formidable achievement for a company who had only five years of real experience of electricity behind them.

The main winders and electrical equipment were erected before the actual sinking commenced in June 1910. Work was severely impeded by the presence of large and unexpected quantities of water. But for the installation of powerful, high lift pumps operating continuously for long periods,[50] many would have doubted the ultimate success of the venture. To ensure continuity of supply to pumps in high risk areas, duplicate double-wound armoured cables were installed.[51]

The two electric winders were supplied by Siemens Brothers Dynamo Works. The motors, directly coupled to the winder, operated in the Ilgner principle (see chapter 6) where the incoming 10,000 V supply was converted to 600 V dc.[52] Although each winder was capable of developing a maximum of 4,300 bhp when winding six tons of coal per minute from 730 yards, equalisation ensured that the maximum energy drawn from the supply did not exceed 1,750 hp.

During the sinking the induction motors were supplied directly with 3,000 V, rather than 10,000 V.[53] One Ilgner set drove both winders during the sinking stages. This was possible because each convertor had two dc generators and the output of each was simply connected to the winder motors.

8.8.5 Both systems — an overall appraisal

By 1914 the PDSC 'empire' covered more than 16,000 acres.[54] Initially, electrical developments were confined to the Aberdare valley and the Rhymney valley (Figs. 8.3 and 8.4 show these areas in greater detail).[55]

In the Aberdare valley 10 collieries were supplied by a 3,000 V transmission system (capable of uprating to 10,000 V when demand rose), and in the Rhymney valley five collieries were supplied from a 10,000 V system.[56]

To render the system completely flexible, in the spring of 1915 a 30,000 V tie (to be worked initially at 20,000 V) was added between the Middle Duffryn power station and the new Britannia colliery. The linkage was not a new idea — it was proposed as early as 1910 but had then to be abandoned not on technical grounds but because of difficulty in negotiating wayleaves. Such a link was expected to raise the load factor of the overall system to between 55% and 60%.[57]

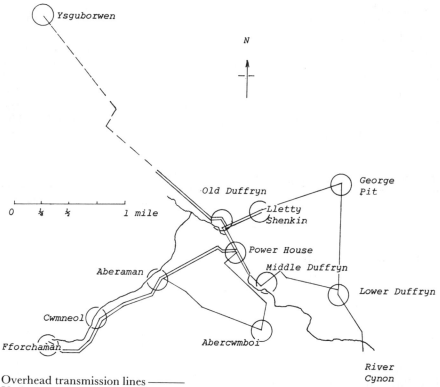

Overhead transmission lines ———
Underground cable – – –

Fig. 8.3   Transmission system for the PDSC collieries in the Aberdare valley

In 1905, Sparks had expressed an opinion:

'that the reduction in the first cost of electrical machinery would result in rapid progress being made in the utilisation of electric power for all colliery work'.[58]

These were prophetic words for by 1914 the PDSC had installed plant rated at 44,800 hp in the Aberdare and Rhymney valleys. Table 8.2 breaks this figure down and demonstrates clearly the developing policy of the PDSC.[59]

Pumps were by far the largest consumer of electricity — a total of 16,610 hp. When steam was the motive power, the main pumping stations for the Aberdare and Rhymney valleys were situated at Middle Duffryn, Elliot Pit and Bargoed. Electricity made a more diversified approach possible and pumps were located at three collieries in the Aberdare valley and three in the Rhymney valley, which combated the 'wet pit' problem better.

By operating the main Rhymney valley pumping station at the Elliot Pit between eight and twelve hours daily and the others, where possible, during the night, a much better load factor was achieved. In the Rhymney valley night pumping, combined with the other colliery requirements, enabled the gas engines to operate at an annual load factor of 72%.[60]

The second largest power user was haulage, with a total of 99 drives, of which 55 were underground. This accounted for more than 10,000 hp. The PDSC preferred the main and tail type, with individual drives up to 250 hp. Single gear reduction was used on the larger haulages where space was not a consideration, and double gear reduction on smaller haulages where the space factor was important. Induction motors were the accepted form of drive underground because of the danger from sparking on dc motors. Rotary convertors provided direct current. The cost of running the haulages was high because of the intermittent nature of their load which increased towards the end of the shift. The best way to reduce costs was to run as many haulages as possible from one generating station.[61]

Table 8.2   Summary of plant installed in PDSC (1914)

	Number of motors	HP
Winders	9	7,530
Pumps — above 100 hp	36	14,490
— below 100 hp	60	2,120
Fans	14	2,720
haulages — above ground	44	3,800
— underground	55	6,370
Screens & elevators	61	1,285
Washery	20	800
Air compressors	9	1,830
Miscellaneous motors used for surface work	227	3,855
Total	535	44,800

Only nine electric winders were installed within the group. Whilst the PDSC policy on the use of electricity was progressive, they saw no need to install new electric winders, which could cost as much as £11,000[62], unless the existing steam-driven ones needed replacing or a new pit was sunk. Those they did install show no real preference between the Ilgner (or one of its variations) or a direct coupled ac motor. The latter type was one of the first installed within the group at Abercwmboi about 1905.

The fan drives, though few in number, had begun to reflect modern colliery practice in replacing low speed fans with more efficient high speed ones. They were driven by induction motors either directly or rope driven, depending on the application. Direct drive was favoured where replacing or supplementing existing steam drive operating at full output. A variable resistance in the rotor circuit catered for small variations in speed. If the fans were to be used in the opening out of new workings the quantity of air could only be estimated. Initially rope drives would be installed there, until the full extent of the workings had been realised, when a permanent fixed speed motor would be fitted. Other solutions involved the use of 'cascade' or dc motors. There was an unusual arrangement at the Britannia colliery where they had a dc motor, but the rotary converter also drove the compressor motor.[63]

The dangers of the use of electricity underground still remained a matter of the utmost concern for the PDSC, but they were certainly not alone in this. They continued to use compressed air as the sole motive power for coal cutters, conveyors, drills and haulages located in 'sensitive' areas. Wherever practicable, the air was supplied by local, isolated, electrically-driven compressors. Many of these were in far from ideal environments and required frequent attention and maintenance. To ensure that production was maintained during such maintenance, the main compressor, invariably located in the winder house, was connected to provide a back-up supply. Although this system was expensive initially, it ensured continuity of production and reduced the need for spares and reserve plant.

Because of the intermittent demand for air, it was often uneconomical to run compressors at full output. To reduce the waste, the PDSC installed variable speed, dc drives in a number of their collieries. The doubling up of the rotary convertor at the Britannia colliery, to drive both the compressor and the ventilation fan, resulted in a very satisfactory and efficient drive.

The generated output for 1914 was about 50 million units.[64] The power factor of the Rhymney valley varied between 0·7 and 0·8, whilst the Aberdare valley gave slightly better returns between 0·7 and 0·85.[65] The overall load factor was about 37%. The generating plant that catered for this demand is listed as follows:[66]

- *Aberdare valley*

  Middle Duffryn power station
  1 × 5,000 kW Escher Wyss–Siemens high pressure turbo-alternator running at 1,500 rpm
  2 × 2,000 kW Escher Wyss–Westinghouse high pressure turbo-alternator running at 1,500 rpm
  1 ×2,000 kW Curtis high pressure turbo-alternator running at 1,500 rpm

- *Rhymney valley*
Penallta
2 × 3,000 kW mixed pressure Westinghouse turbo-alternators running at
1,500 rpm

Bargoed
2 ×1,600 kW Nuremburg gas engines driving AEG flywheel alternators
running at 100 rpm
1 × 800 kW Nuremburg gas engines driving AEG flywheel alternators
running at 100 rpm
1 ×2,000 kW Fraser & Chalmer/Siemens mixed pressure turbo-alternator
running at 3,000 rpm

Elliot Pit
2 × 500 kW Fraser & Chalmers exhaust steam turbines driving Dick Kerr
generators running at 1,500 rpm

This list, with its variety of equipment types and manufacture, well illustrates the
expansion over the years, particularly when contrasted with the early equipment
installed. For example, the generating equipment in the Middle Duffryn was low
speed reciprocating engines driving flywheel generators which produced a total
of 3,000 kW.[67]
The PDSC, whenever possible, utilised the exhaust steam from the winders
and other large equipment to drive their generators. Whilst this principle is
sound in theory, sufficient steam must be available to ensure that the prime
mover is maintained at a constant speed. In this respect the PDSC had mixed
fortunes.[68] At the Penallta power station, the two 3,000 kW mixed pressure
turbo-alternators were easily powered off the exhaust steam from the two winders
and a compressor. The plant ran so reliably that they had no reservations in
running it in parallel with the gas engine station at Bargoed. Close speed
regulation was essential to avoid overloading the gas engine each time the
Penallta plant was changed from high to low pressure steam. It was a different
matter at Elliot Pit, where steam availability limited generation to the day shift.
Here the 500 kW Dick Kerr alternators (which were not the original) had new
driving-ends fitted in 1909 in an attempt to improve the situation. Further
modifications changed the design of the exhaust steam Rateau turbines to Fraser
& Chalmers, but these alterations resulted in only limited success.
As a larger more advanced plant was installed, many of the original
reciprocating sets that became redundant were not discarded but used for
pumping and ventilation during emergencies.[69] Such plant was located at:

Aberaman : 1,500 kW
Bargoed : 750 kW
Penallta : 750 kW
Elliot Pit : 750 kW

The equipment was retained for another reason: as an insurance against
increasing labour problems.
Although the sophistication and flexibility of high voltage transmission
systems had rapidly improved over the years, as far as collieries were concerned it

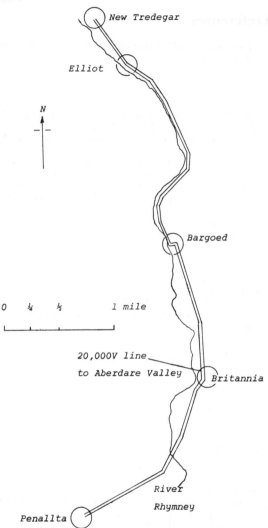

Fig. 8.4 Transmission system for the PDSC collieries in the Rhymney valley

was only to one purpose: to provide a reliable and adequate source of energy that would enable them to produce more coal at less cost.

In 1914 PDSC raised approximately four million tons of coal, which was 7·4% of the total South Wales output.[70]

8.9 Notes and references

1 SPARKS, C.P.: 'Electricity applied to mining', *Journal of the Institution of Electrical Engineers*, 1915, Vol. 53, p. 391
2 *ibid.*
3 *ibid.*, p. 430
4 *ibid.*
5 *ibid.*, p. 437
6 BYATT, I.C.R.: *'The British Electrical Industry 1875–1914*, Oxford, 1979, p. 115
7 CHANT, J.W.: *'The Electricity Supply Acts'*, 1948, p. ix
8 WILSON, J.F.: *'Ferranti and the British electrical industry, 1864–1930'*, Manchester, 1988, p. 19
9 HANNAH, L.: *'Electricity before nationalisation'*, London, 1979, p. 7. Also POULTER, J.D.: *'An early history of electricity supply'*, Peter Peregrinus, Stevenage, 1986, p. 91
10 POULTER, J.D.: *An early history of electricity supply'*, Stevenage, 1986, p. 91
11 BOWERS, B.: *A history of electric light and power'*, Peter Peregrinus, Stevenage, 1982, p. 158
12 BYATT, I.C.R.: *op. cit.*, p. 113
13 THOMAS, T.S.: 'Collieries and public supply', *Mining Electrical Engineer*, 1926–27, Vol. 8, p. 276
14 BYATT, I.C.R.: *op. cit.*, pp. 113 and 115
15 MARTIN, J.S.: *'Mines Inspectors Reports'*, 1909, p. 18
16 BYATT, I.C.R.: *op. cit.*, p. 115
17 HANNAH, L.: *op. cit.*, p. 52
18 MERZ, C.H.: 'Power supply and its effect on the industries of the North-East Coast', paper read to the Iron and Steel Institute at Middlesbrough and reprinted in *The Electrician*, 2 October, 1908, p. 959
19 *ibid.*, p. 959
20 HANNAH, L.: *op. cit.*, p. 31
21 MERZ, C.H.: *op. cit.*, pp. 959–960
22 HANNAH, L.: *op. cit.*, p. 32
23 MERZ, C.H.: *op. cit.*, pp. 959–960
24 HANNAH, L.: *op. cit.*, p. 32
25 MERZ, C.H.: *op. cit.*, p. 960
26 *ibid.*, p. 961
27 *ibid.*
28 BYATT, I.C.R.: *op. cit.*, p. 121
29 *ibid.*, pp. 113–114
30 JONES, E.: 'The Cardiff Electricity Undertaking 1891–1914'. Paper (unreferenced) given in Cardiff, 1914, p. 95
31 SPARKS, C.P.: *op. cit.*, p. 430
32 *'The Powell Duffryn Steam Coal Company Limited 1864–1914'*. Published by the PDSC to commemorate its Golden Jubilee, p. 7
33 MORRIS, J.H., and WILLIAMS, L.J.: *'The South Wales coal industry 1841–1875'*, Cardiff, 1958, p. 162
34 *ibid.*, pp. 69–70
35 Charles Pratt Sparks, CBE (1866–1940) had an illustrious career in electrical engineering, and in obituary notices [see *J.IEE*, 1941, Vol. 88, Pt. 1, p. 318] he was acknowledged as 'one of the band of pioneers . . . who, under the leadership of Kelvin, Ferranti, John Hopkinson and many others, laid the foundations of electricity supply as we know it today.' His approach to the PDSC installation was predictable in view of his work on high voltage ac distribution with Ferranti.
   Sparks also had the rare distinction of being President of the IEE twice (1915 and 1916). READER, W.J.: *A History of the Instutition of Electrical Engineers 1871–1971'*, 1987, p. 271
36 SPARKS, C.P.: *op. cit.*, p. 397
37 DAY, B.J.: 'Visit to Powell Duffryn Collieries and Works in the Aberdare Valley', *Proceedings of the South Wales Institution of Engineers*, 1904–1906, Vol. 24, pp. 560–579
38 *ibid.*, p. 562

39 SPARKS, C.P.: *op. cit.*, pp. 390–391
40 DAY, B.J.: *op. cit.*, p. 579
41 *ibid.*, p. 561
42 *ibid.*, p. 571
43 *ibid.*, p. 560
44 MARTIN, J.S.: *MIR*, 1904, p. 22
45 *ibid.*, 1906, p. 24
46 *'Powell Duffryn Steam Coal Company' op. cit.*, p. 68
47 SPARKS, P.C.: *op. cit.*, p. 396
48 *ibid.*, p. 390
49 *'Powell Duffryn Steam Coal Company'*, *op. cit.*, p. 26, states: 'This [Britannia] is the first colliery in the country to be entirely sunk and worked by electricity generated by the colliery company itself.'
    TOMOS, D.: *'The South Wales Story of the Association of Mining Electrical and Mechanical Engineers'*, Cardiff 1960, p. 16, states that the Maritime Colliery, Pontypridd, Glamorgan, was the first all-electric pit of any size to be operated, but not sunk electrically. This colliery, re-sunk and re-equipped in 1908 (see chapter 6), purchased electricity from the SWEP at 0·5d. per unit.
    Hamsterley colliery, Co. Durham, was also sunk in 1908 using electric pumps and winding. This, however, was a small pit.
50 SPARKS, C.P.: *op. cit.*, p. 390
51 *op. cit.*, p. 398
52 *ibid.*, pp. 400–409
53 Accounts differ as to the reason why the winders were run at 3,000 V as opposed to 10,000 V. TOMOS, D.: *op. cit.*, p. 16 suggests that this was sufficient as the winders would have taken a small load during sinking. Despite the attraction and viability of such a statement, SPARKS (*op. cit.*, p. 408), without giving reasons, is very clear that 'the 10,000 V was not available at the commencement of the sinking'.
54 SPARKS, C.P.: *op. cit.*, p. 389. Fig. 8.5 based on 'flysheet map' in *'Powell Duffryn Steam Coal Company'*, *op. cit.*
55 SPARKS, P.: *op. cit.*, pp. 390–391
56 *ibid.*, pp. 397–398
57 *ibid.*, pp. 397–398
58 *ibid.*, p. 389
59 *ibid.*, p. 423
60 *ibid.*, pp. 412–413 & 396
61 *ibid.*, p. 416
62 See pp. 124–125 for breakdown of costs relating to steam and electrically driven winders at the Great Western's Maritime Colliery
63 SPARKS, C.P.: *op. cit.*, pp. 414–415
64 *ibid.*, p. 389
65 *ibid.*, p. 404
66 *ibid.*, p. 393
67 *ibid.*
68 *ibid.*, p. 394
69 *ibid.*, p. 393
70 *ibid.*, p. 389

Chapter 9
Colliery electrification — in retrospect

9.1 The dawn of a new era

The rapid expansion of industrial activity throughout the western world during the nineteenth century was founded on two materials: coal and iron. At the beginning of that century, Britain had a commanding lead in coal production over every other nation, with 80% of world production coming from the British coalfield. Against this background, one can understand George Stephenson's suggestion that the Lord Chancellor should sit on a sack of coal in the Palace of Westminster, rather than the sack of wool, which was out of date as a symbol of Britain's commercial power base.[1]

As the century developed, the demand for coal in Britain mushroomed as new industries were created, each demanding more and more fuel. As industry expanded, so did the demand for labour and hence the need for coal as a household fuel. Cheap and plentiful supplies of iron meant that machinery and hence manufactured goods could be produced more cheaply.

Essential to nineteenth century industrial and commercial expansion was the development of railways. Following the steam-powered locomotive by Richard Trevithick in 1804, a small number of rack railways were laid down, notably at Leeds in 1812, and shortly afterwards at Newcastle upon Tyne. However, the development of the now conventional, smooth-wheel drive engine, both by William Hedley and George Stephenson, for use at collieries on Tyneside led directly to the birth of the railway era.

In 1825, as little as 26 miles of railway track had been laid down in Britain. In 1880, this had increased to 15,563 miles and by 1913 there were 20,266 miles of track in use.[2] Opening up these inland trade routes stimulated the demand for goods and created new markets. By 1860, some 88,400,000 tons of freight were being moved by rail. By 1913, this total had risen to 364,400,000 tons.[3]

A class of new industries arose that depended directly upon coal itself. One of the first and most obvious was the manufacture of town gas, and its production processes yielded a number of important by-products, including ammonia and coal-tar. Further distillation of coal-tar produced creosote, benzene and naphthalene. It could be argued that the most important development from the gas production process was the extraction of aniline, for around this chemical a whole dyestuffs industry blossomed. By the end of the nineteenth century this industry had developed into the manufacture of other chemicals, notably pharmaceuticals and explosives.

The second half of the nineteenth century is perhaps best remembered for the development of heavy engineering, including armaments. Following the introduction of the Bessemer Convertor (1870) and the Siemens open hearth furnace in the next decade, the world, according to Alfred Krupp, moved into the 'steel age'. In 1871, steel production in Britain was around 329,000 tons per annum — some 35% of world production. By 1913 British production had increased some twenty-fold to 7,835,000 tons, but its share of world production had dropped to 10%. German output increased similarly, but the largest increase was in the USA, where by 1913 over 40% of the world's steel was produced.

The high seas benefited from the advent of cheap steel, with expansion of both the British merchant fleet and the Royal Navy, which opened up and protected the Empire's trade routes. One of the major industries to profit from this increase in overseas trade was textiles. By the early twentieth century, it was said that the Lancashire cotton industry produced for the home market before breakfast and for export the rest of the day.

Thus, by 1914, there was hardly a facet of life in Britain, or other Western countries, that did not rely wholly, or in part, on the continued production of coal. It is against this background that developments in the coal industry should be viewed.

9.2 A false security

Further analysis of Fig. 5.1 and Tables 5.2 and 5.3 relating to the output of the British coalfield reveals that the Victorian prosperity, so soundly based upon coal, could not continue.

Although outputs had increased from 147 million tons in 1880 to more than 287 million tons in 1913, the output per man had actually fallen from an average of 319 tons to 257 tons.[4] Malaise was thus established at an early date.

These ever-increasing outputs were achieved (with the exception of Scotland discussed further in section 9.4.2) without any significant increase in labour-saving devices. Indeed, it would be no exaggeration to say that at the actual coalface, in the greater majority of mines, men were working in conditions that had hardly changed for more than a century. Of the coal produced in 1913 — the year of peak productivity in the British coalfield — 8·2% was mechanically cut.[5] By 1929 this figure had only increased to 28%. By comparison, the United States could boast then 78%, and the Ruhr region of Germany, Britain's main competitor, 91%. Yet in 1913 the percentage of mechanically cut coal in the Ruhr had been only 2%. Clearly Britain had failed to develop at the same rate as other countries, although much of the technology was pioneered in Britain and readily available. Natural conditions could not be held responsible, for they were similar to the Ruhr.

The Reid Report, published in 1945, included an historical survey of British coalmining before and after 1926, and summarised the pre-1926 situation concisely:[6]

(i)   Mining practices throughout Britain were based on long-standing local customs, traditions and methods of mining which differed considerably from coalfield to coalfield, even when conditions were similar. The result was that developments occurred at a rate determined by the initiative and expertise of mining engineers employed by individual companies. This was often very gradual.

(ii)   There were too many small and widely dispersed mines throughout the coalfield. In 1900 Britain had 3,089 mines, with an average output of 73,000 tons.

(iii) The independent character of many of these self-contained mines mitigated against major technological improvements. Managers were not encouraged to visit other installations at home or abroad, although they could learn about technical developments and information disseminated by learned bodies. Small collieries could not afford the luxury of employing technical staff beyond those required for the day-to-day supervision of the colliery, which engaged all their time and energy.

(vi)   Capital resources of the small mineowners were very limited. Forward planning, even on a short-term basis, was unusual. Any available capital was used to satisfy the immediate demand for coal.

Robert Nelson, the Electrical Inspector of Mines who was intimately involved with the industry during this period, offered another reason: the widespread belief that electricity was responsible for a number of disastrous explosions between 1908–14. (This will be considered later.)

The twentieth century colliery owner inherited a legacy of mines, which, without major restructuring, could not be readily adapted to modern techniques. At least one-half, possibly two-thirds, of the coal mined came from collieries established before 1875.[7] This meant that the better and more accessible seams were nearing exhaustion, leaving the thinner and heavily faulted seams to be worked with increasing difficulty and expense.

Formidable though these factors were, there was really very little pressure on British coalowners to modernise or change. Many were lulled into a false sense of security because they saw no reason why the ever-increasing demand for coal should not continue. Competition from abroad seemed no real threat, and, should it ever become so, it would be countered by the method adopted against internal competition — employing more miners. Labour was cheap and plentiful. The increase in manpower from about 500,000 to 1,250,000 between 1880 and 1913 bears testimony to this.[8]

The outbreak of the First World War did little to improve the situation, despite the desperate need for coal. The relationship which evolved between the British coalmining industry and the Government during this period is a study in its own right, its events and time scales beyond the scope of this book. Severe difficulties were created by miners leaving the industry to fight for their King and country (in the first year of the war more than 250,000 men left the industry).[9] To offset this several hundred coal-cutting machines were imported from the United States, but other developments were delayed and machinery was replaced only when absolutely essential. Government took control of the mines, leaving the owners responsible only for day-to-day running. In due course, bitter, protracted labour disputes also rebounded onto the owners.

The problems of the coal industry, as outlined by the Reid Report, were not 'new'. They had been highlighted in 1919, and not for the first time then. The Sankey Report of 1919 produced the following, strongly worded statement:

'the present system of ownership and working in the coal industry stands condemned, and some other system must be substituted for it.'[10]

With half of the members chosen by the Miners' Federation of Great Britain and the promise that whatever was proposed would be made law, the Sankey Commission suggested a major breakthrough in the miners' cause. A number of solutions were put forward, of which nationalisation was one.

Sir Richard Redmayne was careful not to recommend nationalisation in his evidence, but went to some length to explain how other forms of 'Collective Production' (as he termed it) would result in a wide range of improvements.[11] The report came out heavily in favour of the miners and the principle of public ownership was accepted by the members, albeit by a majority of one. The miners cause seemed assured. It was not to be: the Prime Minister repudiated the Government's pledge. An opportunity to take fuller advantage of one of Britain's most valuable assets was lost.[12]

There is little doubt that the technology existed to bring about large-scale improvements in both production and working conditions. Notwithstanding the established technologies of compressed air and steam, electricity had by 1910 proved itself a viable alternative in many colliery applications. Large colliery companies which had the capital and resources were not slow to see the advantages electricity offered. New technology would only be adopted if it would produce larger quantities of coal faster and more economically. Electricity, as a 'new' technology, had to gain acceptance.

9.3 Safety considerations

In mining, perhaps more than in any nineteenth-century industry, a safety policy had to be rigorously pursued, particularly in relation to that most lethal hazard, firedamp. The widespread fear of this killer gas was the most serious factor retarding the early development of electrical engineering in mines. During the period under consideration, the presence of methane in a mine presented the biggest potential hazard to life, but it was not the largest single cause of death in mining.

W.H. Preece drew attention to this when, at a meeting of the Royal Institution in 1888, he told his audience that in 1886 there had been 953 fatalities and, of these, 129 (13·5%) were due to explosions, whilst 461 (48%) resulted from falls of roof or walls. But, as he pointed out: Explosions attract immense attention from publicity and by their appalling suddenness and magnitude'.[13]

In fiery mines, the danger was ever present and they had continually to guard against disasters such as Abercarn in 1878, when 268 lives were lost, and at Seaham in 1880, when 164 miners were killed. Preece could have added that the relatively low percentage of fatalities from explosions in 1886 resulted from diligent observance of a number of safety regulations, otherwise the number of deaths could well have been much greater.

Within a decade of its introduction, electrical engineering was generally accepted in mining (although by no means universally) as having advantages over other forms of power transmission as regards the ease of installation, maintenance and efficiency of transmission. Two main problems confronted those who wished to press for its wider application. The first, that of 'technical innovation', was the need for equipment specially suited to the difficult conditions found in mining. The second, 'commercial exploitation', was to encourage mine owners to use the equipment and see that the new systems were fully developed.

The mine owner had to balance his technical responsibility to run the mine efficiently against his legal responsibility to prevent accidents. Acts of Parliament and Home Office Regulations specified requirements for electrical equipment and installations and many mine owners erred on the side of safety. This was a further reason for the slow adoption of electricity in collieries.

Although the use of electricity in mines was mentioned as early 1852, another thirty years were to elapse before an electric motor was first used underground in a British mine. This was at the Trafalgar Colliery in the Forest of Dean (see chapter 3). Meanwhile, electricity was successfully used in such applications as shot-firing, signalling and, latterly, lighting, Nelson believed that one reason for this protracted period was that mining, being 'already an old industry with fixed traditions and well established practices', presented a formidable barrier to electrical engineers. It required an extraordinary patience and persuasion for electrical engineers to effect any change, particularly as they were offering a 'new, untried and even "mysterious" agent'.[14]

Despite the success of the Trafalgar Colliery installation, the growth of electricity in mining was very slow. By 1900, long after electricity had lost its 'novelty', there was probably no more than 10,000 hp of installed electricity in the British coalfield. At least this allowed time for development, as manufacturers sought to understand machine design better and to increase the reliability and safety of electrical apparatus for use in hostile environments which were not fully understood. Nelson often used to say that:

> 'Electrical and mining engineers, had, however, still to do a great deal by way of educating each other, and themselves, before they evolved practical and satisfactory solutions of all the problems presented by the distribution and use of electricity under mining conditions.'[15]

They had embarked on a very demanding course of learning — not all were successful.

9.4 Breaking down the barriers

9.4.1 Lighting
The possibility of using the carbon filament lamp was recognised very soon after it was invented, encouraged by the Royal Commission on Accidents in Mines. A number of designers quickly produced prototype safety lamps, as well as hand-held and fixed dynamo-fed lanterns. In the early stages, as Snell recorded (see chapter 2), battery technology limited the development of electric safety lamps, but after about 1900 this argument no longer held good. By 1911 only 4,298 electric safety lamps were reported in use in this country; this represents just over 0·5% of the total number of safety lamps in use that year. This figure had jumped from the previous year's total of 2,055, probably thanks to the findings of the inquiry after the Whitehaven and Hulton Colliery disasters.[16]

From 1910 to 1914, the numbers of lamps in use in Britain leaped from 2,055 to 75,707, nearly 37-fold, but even then 90% of the lamps in use relied upon one of the traditional forms of safety lamp (Table 2.2). This extremely slow adoption of electric safety lamps was considered serious enough for the Secretary of State for the Home Department to give his backing to a competition for the design that complied best with a set list of requirements. Sir Alfred Markham, MP, a colliery proprietor, offered £1,000 in prize money. The CAEG [*sic*] lamp won £500 and ten others were awarded £50 each.[17]

Lighting fed by fixed wiring, both above and below ground, was not subject to all the restrictions applied to the battery operated lamps. As the number of installations grew, colliery management and staff gained valuable experience with this new medium of power. With the simple equipment of the day, problems appeared to be few; in a number of cases it is recorded that extra or specialist staff were not needed to maintain the new equipment.[18]

The technological advances in this field were less spectacular than those associated with some power applications. Nonetheless, lighting installations played an important role in the development and early acceptances of electrical engineering in mining. During the 1880s there were a number of lighting installations in fiery mines, such as the Risca Colliery in South Wales and the Leycett Colliery of the Madeley Coal & Iron Company.

In some cases this early work was carried out to a standard below what was recommended at the time and would certainly not have been allowed twenty years later. One example, mentioned earlier, was the use of old wire ropes, rails, gas and water pipes as supply cables at Cannock Chase Collieries. Such an installation was clearly at variance with both the spirit and the letter of the 'Rules and Regulations for the Prevention of Fire Risk arising from Electric Lighting', issued in 1882 by the Society of Telegraph Engineers and Electricians.[19] Insurance companies, of which the principal one was the Phoenix Insurance Company, had similar rules. They were intended to give guidance and instruction on the safe installation of electrical plant to protect both life and property. The regulations recommended that a skilled and experienced electrician supervise the works, pointing out that the chief danger arose from 'ignorance and inexperience on the part of those who supply and fit up the requisite plant.'

Regulation No. 10 stated that gas and water pipes 'should in no case be allowed' as conductors, and Regulation No. 13 said that it was: 'most essential that the joints [in conductors] should be electrically and mechanically perfect'.

Ensuring that the joints between the bundles of wire ropes, rails and pipes were sound would be far from easy and, where underground piped service mains were used, electrolytic action could occur with resultant corrosion in joints and elsewhere.

Regulation No. 15 directs that all wires should be 'efficiently insulated'. It would be difficult to argue that wire ropes wrapped in old brattice cloth or tarpaulin satisfied this requirement. There were a number of acceptable cables on the market, the most appropriate perhaps being Callender's vulcanised bitumen type which became very popular in mining applications.[20]

9.4.2 General mechanisation

In mining, innovation is indispensable if the industry is to be successful. In any extractive industry, a diminishing resource is being pursued and alternative sources must always be sought, which is expensive. In mining, the cost of extraction is never cheaper than at the start of the operation. When a mine is first sunk, the uppermost productive seams are worked first and the coal nearest the shaft is extracted. Thereafter, the retreating face must be followed, with longer inward and outward journeys for both men and the coal haulage system, and these longer roads necessitate greater maintenance and increased costs. To exploit deeper seams, plant of greater capacity will be required for winding, haulage, pumping and ventilation; once again increased costs will be incurred.[21]

The slow take-up rate of coal cutting machines by mine owners in Great Britain was highlighted by the HM Inspector of Mines for the No. 8 (Midland) District in his annual report for 1904:

> 'in future many of the thinner seams must be undercut by the machines if they are to be worked economically. Holing by hand necessitates a considerable waste of coal due to slotting, so that the holer may reach under the coal seam. In thin seams, the amount of labour required for the small output of coal obtained by hand-holing renders it necessary that the under-cutting should be done by machinery with the least waste of coal.'

Buxton claims that, within the British coalfield, Scotland led the way before the First World War in the application of new technology, with a policy of exploiting diminishing reserves to the full.[22] In this, the Scottish mine owners had the full backing of the miners' union. Their position was quite clearly stated by Robert Smillie, in evidence to the second Committee on the Use of Electricity in Mines, when he said that coal cutting machines were:

> 'absolutely necessary for the development of our output in the very thin seams . . . many seams would be idle today had it not been for the use of coal cutting machines.[23]

As can be seen from Tables 5.2 and 5.3, even before 1914 there was a predominance of electrically powered coal-cutting machines in use in Scottish mines. The number of compressed air-powered units was also significant. In the early days electrically powered plant was evidently less reliable than the pneumatic counterparts, and, especially in country areas, spare parts and skilled men were not always available to repair electrical equipment.[24]

In contrast to Scotland, the steam coal pits in South Wales were very gassy and the majority had bad roofs, and these are cited as reasons for the Welsh reluctance to use electrically powered machines at the coal face.[25]

Fig 9.1 shows that from 1900 to 1914 there was a slow but steady increase in the number of coal cutting machines used in British pits. These machines were particularly suited to the conditions in Scottish mines. In 1906, 4% of Britain's total national output was mined by machines, but 8% of Scotland's. By 1914, the Scottish figure had risen to 23·7% (compared with 9% for the United Kingdom as a whole), with electrically operated coal cutters contributing 82% of that total.

This trend to increased mechanisation at the coal face continued in Scottish pits and by 1924 the percentage of machine-cut coal had risen to 47% of the total production, with 1,518 electric and 100 compressed air coal cutters in use. By comparison, the percentage output cut by machine for the whole of the United Kingdom was 19%, with 2,745 electric cutters out of a total of 6,159.[26] For the same year, 70% of coal mined in the USA was machine-cut.[27]

Criticism of the slow adoption of electricity by many mine owners was justified,

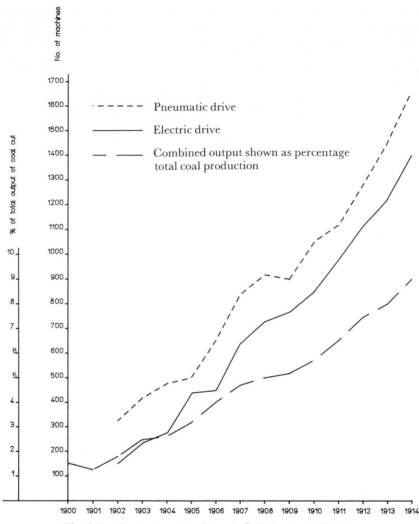

Fig. 9.1   Coal cutting machines in British pits, 1900–14

Data taken from HM Inspector of Mines Reports.
No individual figures for types of machines were given for 1901, the combined total was given as 345.
In 1901 no returns were provided by Newcastle and Midland Districts, therefore, HM Inspector quoted the previous year's returns.

and similar criticism could be made of British industry generally, but that adoption was certainly not helped by gas and difficult physical conditions. It is not surprising that solutions were not rapidly forthcoming and that new equipment was viewed with some scepticism.

L.B. Atkinson tackled many of the varied problems in the early days. His academic training is clearly in evidence, with a sound approach not only to electrical, but to mechanical problems. Atkinson extended and applied the principle, first put forward by Davy, for cooling hot and expanding gas by the use of wide, close fitting flanges, extending the area of contact for the gas by incorporating additional, close fitting, metal surfaces (patent No. 12,676:1887). This system was used extensively later.

A common feature of early, electrically-powered coal cutters was the noise of the gearing, and this was blamed for a number of deaths from roof falls as the miners could not hear the warning sounds preceding a collapse.[28] One way of reducing the machine noise was to run the gearing in a bath of oil. The Atkinson & Goolden machine, as early as 1889 (patent No. 14,041:1889), employed such an oil bath.[29] Development of the Atkinson & Goolden bar coal cutting machine brought together a group of young men of outstanding talent — the Atkinson brothers, F. Hurd jnr. and Hans Renold and it was recognised that this type of machine stood more chance of success than its only competitor at the time, the disc cutter. The electrically-driven chain cutter followed a little later.

The electric drive of the bar type machine did not require such a large starting torque as the disc, nor was it so liable to heavy overloads resulting from a collapse of the coal face. In the Atkinson & Goolden machine, the torque requirement was further reduced by re-designing the cutter bar. The modified characteristics of the bar machine led to its success in: 'association with the comparatively low power and crude design of the early electric motor.'[30]

9.5 ac and dc installations

The patenting of the three-phase induction motor by Tesla in 1888 marked a major advance in electrical engineering and mining, although early difficulties, including the motor's inability to develop adequate torque, ensured that its impact was far from immediate. The Ackton Hall colliery in Yorkshire is generally acknowledged as the first British colliery to use an ac motor underground in 1898.[31]

The use of ac in the British coalfield spread slowly, with the majority of applications concentrated in the North East, Yorkshire and South Wales where substantial public supply authorities' lines existed. These networks could cater for heavy loads, such as winders.

Collieries on the Continent and in the United States were less hesitant about ac and almost from the outset engineered large and efficient systems around it. The British reluctance was to some extent born out of timing and costs. In this respect the pioneering companies which had installed electricity on a large scale in the 1890s were victims of their own progress because, whilst they clearly recognised the benefits of electricity, they had bought dc, the only proven technology then available. Having made this commitment and irrespective of subsequent improvements, colliery owners were not going to incur heavy and, in their view, unnecessary expense by scrapping relatively new equipment and buying the ac equivalent. Understandably, they wanted a return on their investment, but there can be no doubt that their unwillingness to scrap 'outdated' equipment and conservative attitude towards technical developments ensured that Britain lagged behind foreign competition.

The superiority of foreign, especially German, technical education has been suggested as a further factor. John Perry had little doubt about this when, speaking as President of the IEE in 1900, he said: 'Must we, . . . always depend upon our Uitlander population,' and, then more specifically stressed the dependence on: 'Fleming and German, Hollander, Huguenot and Hebrew, for the development of our natural resources.'[32]

The 1902 Inquiry was quick to point out the advantages of electricity:

> 'During the hearing of the evidence, one point in particular impressed us, namely the very large part that electricity is likely to play in mines in the future.'[33]

In respect of ac, they acknowledged the slow start but were optimistic for the future:

> 'The use of the three phase system has been somewhat restricted in this country, but it seems probable that in the near future its employment will be largely extended . . . In our opinion this system deserves the careful attention of all who intend to install electricity in mines.'[34]

To support these assertions, they drew attention to the main advantage of ac motors, the absence of commutators and hence of sparking:

> 'It is evident that for use in fiery mines a motor without parts liable to sparking is preferable to one with such parts, no matter how well designed the boxing-in may be.'[35]

The transition from ac to dc was far from uniform. Many collieries operated mixed systems. This was not as confusing as it sounds and was a logical progression for those who had earlier installed dc. When the time for expansion arose, ac was often chosen and the two systems coexisted until dc completed its useful life. A large Derbyshire colliery (unnamed) was cited in 1906 as an example where the original dc plant was retained for lighting and new ac equipment installed for underground haulage. This colliery showed no lack of initiative (and perhaps a good reason for not discarding dc), because, in addition to its own lighting requirements, it supplied electricity for lighting to houses, shops and a hotel in a nearby village![36]

This shift from dc to ac, correctly predicted by the 1902 Inquiry, was the biggest single change in colliery electrification, particularly underground where the ac motor offered a solution to the inherent dangers of sparking in fiery mines. Such was the magnitude of change that nothing less than a thorough revision of the existing 1905 rules, which were effectively based on experience in lighting and dc applications, was necessary (see chapter 7).

The 1909 Inquiry confirmed the earlier Inquiry's attitude towards the use of electricity in mines, stating at the beginning of their report that such developments were 'in the right direction'. Provided that electricity was used properly, it 'cannot be regarded as inherently unsafe'.[37]

At this time all sparks were regarded as inherently dangerous — wisely so in the absence of definable limits. If engineers had known that sparks from dc were twenty-five times more dangerous than ac (50 Hz at 500 V),[38] then the transition might have been hastened. Future work by Professor W.M. Thornton and others defined precise conditions relating to the ignition of methane: the higher the frequency the harder it was to obtain break-flash ignition. At 100 Hz (again at 500 V) the safety factor between ac and dc increased fifty-five times. On this basis alone, serious suggestions were made in the early 1920s that the frequency to underground equipment should be increased and, more particularly, that to lighting where the poor level at the coalface mitigated against optimum work output.[39] Commercial and other reasons weighed too heavily; most collieries retained the 50 Hz standard.

Regrettably figures are not available from before 1927 to show the allocation of ac and dc, but in that year ac accounted for 80% of the horsepower installed.[40]

Although the ac system installed by the PDSC was rightly regarded as a 'model' by 1913, the effectiveness of integrated networks and a central generating station, albeit using dc, had been demonstrated as early as 1895 in the USA. Here the Youghiogheny River Coal Company, Pennsylvania, had installed a central station, capable of producing 400 kW at 500 V dc, to a number of drift mines within a three to four mile radius. Power was supplied to a wide variety of electrical equipment, including coal cutters, locomotives, pumps, fans and screens.[41]

9.6 Continuing safety concerns

As electrical equipment went ever nearer to the actual coalface, with electric coal cutters and conveyors now very much a reality, fears based on genuine concern or mere ignorance reached a new intensity. Trade Union backing was given to the use of coal cutters, in Scotland at least, by Robert Smillie, yet he was not in favour of using electric coal cutting machines in fiery mines.[42] Arthur B. Markham said he would prohibit the use of electricity in *all* fiery mines,[43] and quoted the strict rules then in force in the state of Pennsylvania.[44] Markham was not against electricity *per se*, indeed he claimed to have more electricity in the companies that he was connected with than anyone else,[45] but he thought that stronger rules should be enforced in fiery mines.[46] Thus by 1903, powerful voices were demanding strict rules to curtail the use of electrical equipment in fiery mines. As late as 1911, the Derbyshire Miners Association passed a resolution calling for the total exclusion of electricity from mines except for lighting,[47] because the increasing use of electricity had coincided with a series of disastrous explosions, although there was no proof that electricity was to blame. The real villain in these explosions was later identified as coal dust. Robert Nelson was firmly convinced that, had the coal dust danger not been reduced, then 'the growth of the use of electricity in coal mines would have been seriously hampered'.[48] There is little doubt that he was right.

Over this period the number of electrical fatalities in any one year, except for 1914, never exceeded 1·54% of the total number of accidental deaths in British collieries. Of this figure, almost half the fatalities occurred on the surface.[49] There was more electric power (in terms of horsepower) underground, with the pumps and haulages, but more equipment was located on the surface, which explains the higher number of fatal accidents above ground. Nevertheless, contemporary reports indicate that the standard of electrical equipment installed underground was far from satisfactory. Robert Nelson made some pertinent remarks:

> 'the standard of safety underground (apart from such special circumstances as the possible presence of explosive gas) is

undoubtedly much lower than in ordinary surface work, and also much lower, I have no hesitation in saying, that it might be made without added cost by the exercise of a little foresight.'

It was not uncommon to find colliery switchgear, for example, not metal-clad. If it was, there was very little to suggest that it had been designed with underground usage in mind. 'A square peg in a round hole' was, in Nelson's view, an accurate description of much of the electrical equipment installed underground.[50]

The lack of suitable equipment was not confined to British manufacturers. Between 1905 and 1907 'certain' (unspecified) American coal cutters were introduced into a number of British mines. The use of coal cutters, generally, was a vexed question and unfortunately the introduction of these particular models was ill-timed and exacerbated the situation. One engineer, in later reminiscences, summed up the American experience as having 'disastrous results'. The Americans had seen and attempted to secure a good market, but they 'did not know the conditions in our [British] mines thoroughly' and the machines had 'inferior insulation' and 'were not fitted with enclosed gastight-type motors'.[51] Such comments provided ammunition for critics who argued that fatality figures would have been much higher if the use of electricity were more widespread, particularly underground.

Douglas Hay, Professor of Mining at Sheffield University, delivered a lecture at the Sixth International Mining Exhibition in London in 1923, on the theme of electrical developments in coal mining with special reference to safety. He commented that the low fatality rates (an average of 0·95%) over the decade 1912–22, was due to the:

> 'natural caution with which colliery managements have tempered their enterprise, together with the willing co-operation of manufacturers in endeavouring to meet requirements, and the great advance in scientific knowledge and thought in recent years.[52]

Certainly there had been an increase in scientific knowledge and an undoubted improvement in the construction and installation of electrical equipment after 1910. Nelson enthusiastically endorsed this in 1926, using his own experience as a yardstick, when he said:

> 'British-made mining electrical apparatus is at the present moment substantially better than German-made apparatus . . . it may be reasonably claimed that we [Great Britain] have established a lead over *all* other countries,[53]

Whilst there is every evidence that 'natural caution' was shown by some colliery managements, particularly in the aftermath of the Senghenydd explosion in 1913, where electricity was widely suspected of causing Britain's worst colliery disaster, a great many were still shackled by attitudes and organisation long established on the coalfield. In 1924, 1,266 out of Britain's 2,855 collieries were still without electrical equipment.[54]

Real improvements depend on precise knowledge. They could, perhaps, have vastly over-engineered equipment but this in itself is still no guarantee of safety. In the absence of controlled test facilities, they just had to wait for an incident, hold an inquiry and then act upon its findings. History clearly demonstrates the cost in life and pain of such a method.

Britain lost an excellent opportunity to establish a solid foundation at a critical point in the use of electricity in mines, when the 1909 Inquiry Committee failed to endorse the suggestion made during the Evidence stage that the Home Office should officially approve certain types or designs of electrical apparatus for use underground. The reasons behind their thinking were discussed in chapter 7, but are worth reiterating. The Committee believed that the manufacturers and colliery managements would develop wider and safer use of electricity faster and better themselves. This was a big responsibility to place on them, particularly when there were many 'grey' areas in knowledge and public opinion was unfavourable. Some manufacturers, such as Reyrolle, did excellent work and by 1911 were producing explosion-proof equipment. H.W. Clothier of A.C. Reyrolle, writing in 1909 on the subject of switchgear for mines, said that the expression 'good enough for a colliery' was one of derisive praise. Largely thanks to his unstinting efforts over the years to improve the standards of electrical equipment underground, this saying came to mean the final word in quality. In Nelson's opinion, Clothier's name could not be omitted from those who had made 'contributions of importance towards safety in coal-mining'.[55]

Unfortunately, not all manufacturers had access to such expertise, nor the facilities to undertake extensive testing, so their incomplete data, even if encouraging, was viewed with a certain amount of circumspection. Collectively a wealth of information existed amongst individuals, learned societies, universities, institutions, manufacturers, inquiries etc: better central co-ordination and control would have resulted in greater benefits. Instead, it was a confusing situation, and one which J. Samuels was quick to criticise in his Presidential address to the South Wales branch of the Association of Mining Electrical and Mining Mechanical Engineers in 1921:

> 'This kind of work is badly needed here [Great Britain], and it would be to our advantage if we followed our rivals [the American Bureau of Mines and the Westphalian Mining Association] in this respect, because it is admitted that much of their success is due to the close co-operation between the technical and industrial authorities in maintaining an efficient research scheme and a readiness to put into practical operation any modification suggested by the discoveries made.[56]

It was not until 1930 that the responsibility of flameproof testing of motors and switches was transferred from Sheffield University to the Mines Department Testing Station, Buxton, Derbyshire, and official certificates were issued.

9.7 A firm base

Whatever the problems facing the mining engineer or electrical designer in the introduction of new techniques, lack of information on what was happening in other districts or other countries was not one of them. Up to 1914 at least, mining was open to such information, even on an international scale.

At the formation of the American Institute of Mining Engineers in 1871, an engineer from Austria and another from France were made honorary members, as well as Lowthian Bell, prominent in British engineering circles. The list of members of the (British) Institution of Mining Engineers contained a number of engineers residing in various countries with active mining.

The formation of the Association of Mining Electrical Engineers also attracted overseas members from as far afield as Canada, South Africa, Australia and India. Their membership helped to emphasise the need for such a body, as did the relatively large numbers joining in the early years.

The founding of the Association, however, was not welcomed in some quarters, as illustrated by the comments of S.Z. de Ferranti, who:

> 'had objected most strongly to their formation . . . [he] felt they came into existence because other societies were not doing their duty'

On the positive side he did admit 'that [the] Association had a great work of usefulness to perform.'[57]

By 1914, a firm base had been established for the application of electrical engineering techniques in mining, albeit on a comparatively small scale, and the dangers associated with its use at the coal face were ever more clearly recognised and understood.

9.8 Notes and references

1 SMILES, S.: *'The Life of the Engineers, George & Robert Stephenson'*, 1899, p. 276
2 MITCHELL, B.R.: *'Abstract of British Historical Statistics'*, Cambridge, 1962, pp. 225–226
3 *ibid.*
4 HALL, T.: *'King Coal, Miners, coal and Britain's industrial future'*, 1981 pp. 21–22
5 BERKOVITCH, I.: *'Coal on the switchback. The Coal Industry since nationalisation'*, 1977. p. 42
6 *'Coal mining. Report of the Technical Advisory Committee'* (the Reid Report), 1945, pp. 3–4
7 HALL, T.: *op. cit.*, p. 22
8 BURTON, A.: *'The Miners'*, 1976, p. 147
9 *ibid.*, p. 163
10 *Royal Commission on Coal*, (The Sankey Report), 1919, Interim Report, Recommendation IX, p. viii
11 *ibid.*, p. xx
12 NOEL, G.: *'The Great Lockout of 1926'*, 1976, p. 46
13 *The Electrician*, 1888, Vol. 20, p. 377
14 NELSON, R.: 'Electricity in mines: A short survey', *Journal of the Institution of Electrical Engineers*, 1926 , p. 1011
15 *ibid.*
16 *'Mines Inspectors Reports'*, 1911, p. 59
17 REDMAYNE, R.A.S.: *'Men, mines and memories'*, 1942, p. 154
18 *Engineer*, 1886, Vol. 72, p. 402; *Electrician*, 1883, Vol. 11, p. 274
19 Since 1888, the Society, renamed the Institution of Electrical Engineers, has continued to publish Electrical Regulations, the latest edition being the 16th, published in 1991
20 HUNTER, V., and HAZELL, J.T.: *'Development of Power Cables'*, 1956, *passim*
21 HULL, E.: *'Our coal resources'*, 1897, p. 110, and BOYD, N.: *'Coal pits and pitmen'*, 1885, p. 246
22 BUXTON, N.K.: *'The economic development of the British Coal Industry'*, 1978, p. 113
23 *'Report and Evidence relating to the Working of the Special Rules for the Use of Electricity in Mines'* (hereafter referred to as the *1909 Inquiry*), 1911, QQ5187 and 5188
24 *ibid.*, Q652
25 *ibid.*, QQ5373 and 5652. Also *'Report of the Departmental Committee on the Use of Electricity in Mines'* (hereafter referred to as the *1902 Report*), 1904, Q3978

26 'Fourth Annual Report of the Secretary for Mines and Annual Report of HM Chief Inspector of Mines for 1924', Table 43, p. 135
27 Statistical Tables relating to British and Foreign Trade & Industry (1924–1930), 1931, Table 9, pp. 32–33
28 WILLIAMS, J.E.: Derbyshire Miner, 1962, p. 477
29 1902 Report, QQ4389 and 4395
30 'M & C Machine Mining', Vol. 3, 1923, p. 100
31 NELSON, R.: 'Electricity in coal mines: A retrospect and a forecast', JIEE, 1939, p. 598
32 READER, W.J.: 'A History of the Institution of Electrical Engineers 1871–1971, 1987, p. 65
33 1902 Report, p. 6
34 ibid.
35 ibid.
36 LUPTON, PARR and PERKIN: 'Electricity as Applied to Mining', 1906, pp. 228–229
37 1909 Inquiry, p. 5
38 THORNTON, W.M.: 'Some researches on the safe use of electricity in coal mines', JIEE, 1924, pp. 486–478
39 ibid. Comments made by C.P. Sparks during the 'Discussion', pp. 491–492
40 NELSON, R.: op. cit., 1939, p. 603
41 GRESLEY, W.S.: 'Central-Station electric coal-mining plant in Pennsylvania, USA'. Proceedings of the Institution of Civil Engineers, 1897–98, Vol. 131, Part 1, pp. 100–102
42 1902 Report: Q3184 (Robert Smillie was, at the time of the Inquiry, the President of the Scottish Miner's Federation. He later became Member of Parliament for Morpeth, in Northumberland, and published his autobiography entitled 'My Life for Labour' in 1924)
43 ibid., Q4273
44 ibid., Q4276
45 ibid., Q4313
46 ibid., Q4278
47 WILLIAMS, J.E.: op. cit., p. 481
48 NELSON, R.: op. cit., 1939, p. 599
49 HAY, D.: 'Recent electrical development in coal mining (with special reference to safety problems)', Mining Electrical Engineer, 1923–24, Vol. 4, p. 55
50 NELSON, R.: op. cit., 1939, p. 598
51 NELSON, R.: op. cit., 1939. Comments made by T.H. Varcoe during the Discussion, p. 622
52 HAY, D.: op. cit., pp. 55–56
53 NELSON, R.: op. cit., 1926, p. 1012
54 ibid., p. 1013
55 NELSON, R.: op. cit., 1939, p. 601
56 SAMUELS, J.: 'The electrification of mines, and the demand for technical training': Presidential Address to the South Wales Branch', MEE, 1921–22, Vol. 2, p. 101
57 Proc AMEE, 1912–13, Vol. 4, p. 21

Chapter 10

Glossary and abbreviations

10.1 Mining terms

Adit   A tunnel driven into a hillside in connection with mineral working, for transport, ventilation or drainage (or all three).

After-damp   A mixture of carbon monoxide and other gases resulting from an explosion or fire.

Bank   The colliery surface; coal-faces are also sometimes called banks (or benks).

Bank work   A primitive system of long-way working practised in Yorkshire, Derbyshire, Nottinghamshire and Leicestershire mainly between mid-eighteenth and mid-nineteenth centuries.

Bell-pit   A shallow shaft, where coal is worked around the pit-bottom until the sides are in danger of collapsing when it is abandoned. Seen in section, it has the general shape of a bell.

Blackdamp   A mixture of nitrogen and carbon dioxide and found generally in coalmines, resulting from the oxidation of coal and timber. Its production may leave insufficient oxygen to support life.

Blower   A violent discharge of firedamp.

Bord and pillar work   North Eastern variant of stall and pillar. The bords were rectangular excavations separated by solid pillars of coal left to support the roof. In the nineteenth century, the pillars were also largely removed in a second working.

Brattice   Where a colliery had only one shaft, this was divided from top to bottom by a vertical wooden partition called a 'brattice'. Air flowed down one segment of the shaft and up the other. Brattices, nowadays usually of cloth, are similarly used in headings to facilitate flow of air.

Chaldron   The principal capacity measure used in the North Eastern coal trade. The Newcastle chaldron was equal to 53 cwt and the London chaldron about 27 cwt.

Chokedamp   A synonym for blackdamp. Many nineteenth-century writers applied this term also to afterdamp, and this has confused some present day historians.

Corf   A hazel basket in which coal was conveyed from the coal face to the surface before the introduction of wheeled tubs and cages. Also spelled **corve**.

Creep   The heaving up of the floor of the roadways underground, otherwise called 'floor lift'.

Crush   Convergence of the strata.

Dip   Coal seams are almost always inclined. Working down-hill is said to be the 'dip' and working up hill to be the 'rise'. The 'strike' of the seam is its level course (at right angles to its inclination or slope).

Downcast   The shaft down which the air flows.

Drift   A heading. A drift mine is one driven from the surface (as opposed to a shaft mine).

Dumb Drift   In furnace ventilation, a tunnel isolated from the furnace carrying the return air into the upcast shaft over the top of the furnace, so as to reduce the risk of explosion.

Engine   In old mining terminology, 'engine' usually meant pumping engine, whether powered by horse, water or steam, so a pit from which water was pumped was an engine pit.

Face   The working face of coal.

Fall   A fall of roof at the coal-face or in a roadway.

Fault   A fracture of a coal seam caused by earth movement; it is called an up-throw (where the seam continues at a higher level) or a down-throw (where it continues at a lower level).

Firedamp   Inflammable gas whose chief constituent is methane. It is emitted from the seam (and also the floor and roof) during working.

Gate (gateway or gateroad)   Underground roadways, a term used mainly in the Midlands.

Getter   A man producing coal at the face. In many cases he was also the filler (*qv.*).

Gin   See whim-gin.

Glenny   A safety lamp; probably a corruption of Clanny, after Dr. W.R. Clanny who invented several lamps, the first in 1813.

Goaf (or gob)   The waste area from which coal has been removed; it is partly filled with small coal and debris.

Gob fire   Spontaneous combustion of small coal in the gob.

Gob road   Gate supported by pack walls for bringing coal from the face in hand-got longwall working.

Hewer   A coal-getter.

Holing   Undercutting the face of coal (hence 'holer', the man carrying out this operation).

Inbye   Towards the coal-face (as opposed to outbye or backbye — away from the face).

Intake   Airway along which fresh air is taken into the workings, as opposed to the 'return' airway carrying foul air away from the workings to the upcast shaft.

Kerf (or kirf)   The cut taken along the bottom of the coal-face originally by a man with a pick (the 'holer') and later by machine. Hand holing produces far more slack than machine holing, so cutting machines were more readily adopted where seams were thin.

Longwall   A system of working coal originating in Shropshire where a number of men work along a coal-face. Roadways are maintained through the 'gob', and there are no pillars with this system.

Main road   The main roadway driven from the pit-bottom through the workings.

Outbye   See inbye.

Packs   In hand-got longwall working, pack walls used to be built to support roadways in the gob. In later practice, packs were also built in strips in the area from which coal had been won, allowing the roof to settle gradually. In modern practice, packs have largely been dispensed with and the roof settles quickly.

Panel working   In bord and pillar, the workings are divided into districts (panels) enclosed within barriers of coal so as to restrict the spread of explosions and to facilitate the working of pillars with a minimum of convergence of the strata. 'Panel' is also loosely and misleadingly applied to longwall districts.

Pit   An ambiguous expression. May mean (*a*) a single shaft, (*b*) a pair of shafts (upcast and downcast) or (*c*) (nowadays) a colliery.

Props   Pieces of wood or steel set vertically to support the roof. Horizontal bars were in later practice set up to the roof over the props.

Rag and chain   An early type of pump operated manually or by animals or water-wheel.

Round coals   Large coal from which the small has been separated.

Royalty   The right to work minerals, usually vested in the freeholder.

Royalty rent (or royalties)   Rent paid by the colliery proprietors for the right to work coal.

Sheave   Pulley wheel.

Staithes   Coal wharves on the Tyne and Wear.

Stall and pillar   Working by driving roadways ('stalls') into an area of coal, leaving pillars to support the roof.

Staple   A shaft connecting one seam with another underground.

Steel mill   A device for providing what was thought to be a safe light. A steel wheel is revolved against a flint from which sparks are emitted.

Stinkdamp (or sulphurdamp)   Sulphurated hydrogen.

Stopping   A wall built to prevent air flowing beyond it.

Stythe   Blackdamp.

Sump   The well at the bottom of a shaft from which water may be pumped.

Tub   A small wagon used for conveying coal.

Tubbing   Shaft lining of wood or iron to hold back the water.

Upcast   A shaft carrying an ascending air current. The top of an upcast shaft is always closed in to prevent air from entering from the surface.

Viewer   The old name for manager.

Whim-gin   A horse driven winding device.

Whimsey   Term applied first to windlasses, then to whim-gins, but by early nineteenth century, usually meaning an atmospheric winding engine. In South Yorkshire atmospheric pumping engines were also called whimseys.

Whole   'Working in the whole' means working virgin coal by the bord and pillar method; as distinct from 'working in the broken' where only the pillars are left to be got.

Winning   Opening out a mine so that it is ready to be used.

Working   The art of producing coal.

10.2 Engineering terms and abbreviations

A   Ampère: a measure of the quantity of electricity flowing in a circuit

ac Alternating current: an electric current which alternately changes its direction of flow

cp Candle-power: out-dated terminology for expressing the light output of a lamp.

dc Direct current: used to be called continuous current. An electric current which flows in one direction.

hp Horsepower

Hz Hertz: used to be called cycles per second. The number of cycles a complete (positive and negative) reversal of an ac which takes place in one second. The standard frequency in Britain is 50 Hz.

kV Kilovolts: 1,000 V

kW Kilowatt: 1,000 W

kWh Kilowatt-hour: often referred to as a Unit. A measure of the consumption of electricity, e.g. a 1 kW electric fire burning for one hour would consume 1 kWh.

Load factor The ratio between actual load and maximum output, and expressed as a percentage. The higher the percentage, the greater the cost effectiveness of the system.

mA Milliampère: one-thousandth of an ampère.

Maximum demand The largest demand (expressed in kW or MW) of electricity in a given period, usually one-half or one hour.

MW Megawatt: 1,000,000 W

Peak load The maximum load taken, even if momentarily.

rpm Revolutions per minute

V Volt: the unit of pressure which causes a current to flow.

w Watt: the unit of power. The work done when a current of 1 A flows through a circuit when the pressure is 1 V. 746 W = 1 hp.

10.3 Abbreviations used in references

10.3.1 Institutions, learned bodies, societies etc.

AIME	American Institute of Mining Engineers
AMEE	Association of Mining Electrical Engineers
AMIE	American Institution of Engineers
ICE	Institution of Civil Engineers
FIME	Federated Institute of Mining Engineers
IEE	Institution of Electrical Engineers
IME	Institution of Mining Engineers
IMechE	Institution of Mechanical Engineers
MEME	Association of Mechanical, Electrical and Mining Engineers
MIE	Midland Institute of Engineers
MGS	Manchester Geological Society
MIMCME	Midland Institute of Mining, Civil & Mechanical Engineers
MIS	Mining Institute of Scotland
NACM	National Association of Colliery Managers
NEIMME	North of England Institute of Mining & Mechanical Engineers
NESCo	Newcastle Electric Supply Company
PDSC	Powell Duffryn Steam Coal Company Ltd.

SSEWIME   South Staffordshire & East Worcestershire Institute of Mining Engineers
SWIE   South Wales Institute of Engineers
TIC   Tredegar Iron & Coal Company

The prefix Transactions/Proceedings/Journal applies accordingly.

10.3.2 Journals, reports etc.

JIEE   Journal of the Institution of Electrical Engineers
MEE   Mining Electrical Engineer (Journal of the Association of Mining Electrical Engineers)
MIR   Reports of the Mines Inspectors

Chapter 11

Chronology

1800	First continuous current from primary cell demonstrated by Volta
1808	Carbon arc lamp demonstrated experimentally
1815	Davy invents safety lamp
1818	Institution of Civil Engineers formed
1831	Faraday's discovery of electro-magnetic induction
1837	First practical electric motor patented by Thomas Davenport
1842	Mines Act prohibiting the employment of female labour or boys under ten years of age underground
1848	Compressed air powered coal cutting machine tried out at Govan Colliery
c. 1850	Steam driven ventilating fans introduced in British pits
1853	First chain coal cutter patented
1856	First bar coal cutter patented
1857–70	Professor Holmes uses steam driven magneto-electric generators to power carbon arc lamps in lighthouses
1860	Pacinotti invents the ring armature
1863	Patent taken out by Ridley & Jones for the first electric coal cutting machine
	Separately excited dynamo patented by Dr. H. Wilde
1866–67	Principle of self-excitation of generators independently developed by C. and S.A. Varley, Dr. Werner Siemens and Sir Charles Wheatstone
1871	Gramme produces his first dynamo
	Society of Telegraph Engineers and Electricians (forerunner of the IEE) formed
1874	Provisional patent taken out by Dr. Wilde for electrically powered coal cutting machine
1878	Swan demonstrates incandescent carbon filament lamp at Newcastle
1879	Edison patents his carbon filament lamp in USA and Great Britain
1879–80	Surface installations out arc lighting at various collieries, including the Maerdy Pit in South Wales
1881	Trials carried out under the auspices of Royal Commission on Accidents in coal mines, using Swan's incandescent lamps at the Pleasley Colliery and Earnock Colliery, Glasgow
	Pit bottom illumination with arc lamps at the Maerdy Pit and Risca Colliery, both in South Wales
	Vulcanised bitumen cable developed by W.O. Callender

First electric power station for both public and general use at Godalming, Surrey

1882   Edison opens his Holborn Viaduct generating station, and another at Pearl Street, New York

First electrical wiring regulations issued by the Society of Telegraph Engineers and Electricians

Hopkinson patents his 3-wire system of dc distribution

Swan introduces his miners' 'portable' electric safety lamp. The separate box of batteries weighed 20 lb. The lamp gave a light output of 2 to 3 candles for 20 hours

First Electric Lighting Act

High voltage dc transmission successfully demonstrated at the Munich Exhibition, Germany

First use of electricity for power purposes underground, at the Trafalgar Colliery, Forest of Dean

1883   Grosvenor Gallery station opened

Thomson–Houston Co. formed, Lynn, Massachusetts

1884   Parsons builds his first steam turbine generator developing 20 kW at 100 V dc at 18,000 rpm

High voltage ac transmission successfully demonstrated at the Turin Exhibition, Italy

1885   Grosvenor Gallery site extended and Ferranti called in for advice: he was appointed Chief Engineer in the following year

Bowers & Blackburn patent their rope-drive coal cutting machine

Siemen Bros. introduce lead covered cable, insulated with rubber or gutta-percha

1886   Swan patents his improved portable miners' safety lamp

Increased lighting at collieries, including a large installation at Cannock Chase

1887   Bowers & Blackburn patent electrically powered bar coal cutting machine and set it to work

Goolden & Atkinson patent 'enclosed motor' intended to be capable of working in fiery mines

Electrically operated pumps replace compressed air pumps at St. John's, Normanton

Atkinson, Ravenshaw & Mori patent their electrically powered bar coal cutting machine

Tesla patents two- and three-phase ac motors

Parsons supply their first 100 kW radial flow turbo-alternator to the Cambridge Electric Lighting Co.

1888   Second Electric Lighting Act (Amendment to the First Act)

1889	C.A. Parsons & Co. formed
1890	Institution of Electrical Engineers incorporated
c. 1890	Extensive electrical installation at Lord Durham's Bunker Hill Colliery, including a 32 hp winding motor
1891	First successful demonstration of three-phase ac transmission and distribution at the Frankfurt am Main Exhibition, Germany
1893	Electric haulage plant at Earnock Colliery in operation
1895	Two 500 V, 150 kW Parsons turbo-generators installed at Ackton Hall Colliery, Yorkshire
1895	Large dc installation with central generating station at the Youghiogheny River Coal Company, Pittsburg, Pennsylvania
1896	British Thomson–Houston, Co. formed in Great Britain
1900	First 1,000 kW turbo-alternator. This was supplied by Parsons to the German city of Elberfield and produced 400 V, single phase ac
	First public three-phase supply from Neptune Bank power station, Newcastle, for the Newcastle & District Electric Light Co.
	Home Office request mine owners to give details of coal cutting equipment used (311 in UK, as compared to 3,907 in the USA)
1902	Inquiry into the Use of Electricity in Mines
	First alternator with rotating field installed by Parsons at the Neptune Bank power station
1905	Large scale three-phase electrification commenced by the Powell Duffryn Steam Coal Company Limited, in South Wales, rivalling the Newcastle & District Electric Supply Co.
	First rules for the Safe Use of Electricity in Mines
	First turbo-alternator to generate at 11,000 V supplied by Parsons to Kent Electric Power Co.
1908	First Electrical Inspector of Mines appointed
1909	Second Inquiry into the Use of Electricity in Mines
	Association of Mining Electrical Engineers formed (now the Institution of Mining Electrical and Mining Mechanical Engineers)
1910	Sinking commenced at the Britannia Pit, South Wales. This was the first pit to be sunk and operated by electricity
	Home Office back competition for safety lamp and mineowners offer £1,000 prize
	2,055 electric safety lamps in use in the British coalfield
1911	4,298 electric safety lamps in use in the British coalfield
1912	75,707 electric safety lamps in use (this compares with 679,572 flame safety lamps
1913	Merz described the Powell Duffryn as having the largest and most complete system of colliery electrification in Britain
	Explosion at Senghenydd Colliery, South Wales: 439 killed
	Peak output of the British Coalfield — 287,430,473 tons
	Revised rules for the Safe Use of Electricity in Mines incorporated into the new Coal Mines Act

Index